CONSUMPTION AND DEVELOPMENT

Also by Jeffrey James

CONSUMER CHOICE IN THE THIRD WORLD
THE ECONOMICS OF NEW TECHNOLOGY IN DEVELOPING COUNTRIES (*with Frances Stewart*)
IMPROVING TRADITIONAL RURAL TECHNOLOGIES
TECHNOLOGY, INSTITUTIONS AND GOVERNMENT POLICIES (*with S. Watanabe*)
THE TECHNOLOGICAL BEHAVIOUR OF PUBLIC ENTERPRISES IN DEVELOPING COUNTRIES
THE TRANSITION OF EGALITARIAN DEVELOPMENT (*with Keith Griffin*)

Consumption and Development

Jeffrey James
Professor of Development Economics
Tilburg University
The Netherlands

150th YEAR
M

St. Martin's Press

First published in Great Britain 1993 by
THE MACMILLAN PRESS LTD
Houndmills, Basingstoke, Hampshire RG21 2XS
and London
Companies and representatives
throughout the world

A catalogue record for this book is available
from the British Library

ISBN 0–333–54696–2

Printed in Great Britain by
Antony Rowe Ltd
Chippenham, Wiltshire

First published in the United States of America 1993 by
Scholarly and Reference Division,
ST. MARTIN'S PRESS, INC.,
175 Fifth Avenue,
New York, N.Y. 10010

ISBN 0–312–08526–5

Library of Congress Cataloging-in-Publication Data
James, Jeffrey.
Consumption and development / Jeffrey James.
p. cm.
Includes index.
ISBN 0–312–08526–5
1. Consumption (Economics)—Developing countries. 2. New products—Developing countries. 3. Technological innovations––Economic aspects—Developing countries. 4. Developing countries––Economic conditions. I. Title.
HC59.72.C6J36 1993
339.4'7'091726—dc20 92–12886
CIP

Contents

Preface

This collection of essays reflects my interest during the past ten years or so in the positive and especially the normative aspects of the relationship between consumption and development. I am grateful to Mr. T.M. Farmiloe for his forbearance with delays in delivering the typescript and to Ms Anne Rafique for the excellent editorial help that her office provided. I am grateful too to Corina Maas for typing portions of the book at short notice and with great accuracy. Throughout the book, the generic use of the term 'he' to connote both 'he' and 'she' has been followed. In following this convention, we in no way wish to imply a bias of any kind.

Jeffrey James
Tilburg 1992

Acknowledgements

Three of the chapters in this book were originally published in journals. They are reprinted here with the kind permission of the publishers. Chapter 2 is reprinted from *Oxford Economic Papers*, Vol. 33 (1), 1981, by permission of the Oxford University Press. (This paper subsequently appeared also in F. Stewart and J. James (eds), *The Economics of New Technology in Developing Countries*, Frances Pinter, 1982.) Chapter 4 is reprinted from *World Development*, Vol. 8 (1), 1980, by permission of Pergamon Journals Ltd, Oxford, UK. Chapter 11 appeared originally as the 'Employment Effects of an Income Redistribution: A Test for Aggregation Bias in the Indian Sugar Processing Industry', *Journal of Development Economics*, Vol. 7, 1980, and is reprinted by permission of Elsevier Science Publishers.

Three other chapters are revised versions of papers published previously. Chapter 5 is a revised version of 'Product Standards in Developing Countries', in F. Stewart and J. James (eds), *The Economics of New Technology in Developing Countries*, Frances Pinter, 1982. Material contained in this version is reprinted with the permission of the publisher. Chapter 6 is a revised version of an article with the same title that appeared in *World Development*, Vol. 15 (4), 1987. Material contained in this version is reprinted by permission of Pergamon Journals Ltd, Oxford, UK. Chapter 10 is a revised version of 'Attitude Change Revisited: Cognitive Dissonance Theory and Development Policy' (with E. Gutkind), *World Development*, Vol. 13 (10/11), Oct./Nov. 1985. Material contained in this version is reprinted by permission of Pergamon Journals Ltd, Oxford, UK.

Chapter 9 is based on material that originally appeared as part of J. James, *X-Efficiency, Advertising and New Products: Applications of the New Theory of Demand to Developing Countries*, D. Phil, Oxford University, 1982. Some of this material was subsequently published in 'The New Household Economics, General X-Efficiency Theory, and Developing Countries', *Journal of Development Studies*, Vol. 19 (4), July 1983.

The author and publishers also wish to thank the following who have kindly given permission for the use of copyright material: *The Review of Economic Studies*, for Figures 1 and 2 from J. Harsanyi, 'The Welfare Economics of Variable Tastes', Vol. 21 (3), 1954; *The*

Columbia Journal of World Business, for Chart 1 from S. Douglas and Y. Wind, 'The Myth of Globalization', Winter 1987, copyright 1987; the *Harvard Business Review*, for exhibit I from J. Hill and R. Still, 'Adapting Products to LDC Tastes', March–April 1984, Copyright © 1984 by the Fellows of Harvard College, all rights reserved.

1 Introduction: Consumption and the Modernization Process

Conventional consumption theory is a part of demand analysis and as such belongs to positive economics, but it also lies at the very heart of welfare economics which is concerned with prescriptive or normative propositions, such as, for example, the 'sovereignty' of the consumer.[1] The principal theme of this book is that in its latter capacity, the value of consumption theory is greatly diminished by its inability to come to grips with – and hence to assess and influence – the changes that form part of what is known as the modernization process.[2]

For one thing, with modernization usually goes an exposure (via urbanization and other factors) to the technical changes in products that are so fundamental a feature of developed economies, but which are for the most part absent from traditional societies. 'In addition to the great "revolutionary" changes, millions of tiny changes also take place'[3] in the developed economies. Yet, for all the obvious importance of new products, the topic is largely ignored in welfare economics, as even a cursory reading of the major works in the field will readily confirm.[4] Less attention still is paid by the authors of these works to the changes in individual values (or fundamental human beliefs) that may often be closely related to technical changes in products. It is not that conventional consumption theory never deals with changes in preferences. The problem is rather that, as Hirschman has pointed out, economists have focused primarily on 'impulsive, uncomplicated, haphazard, publicity-induced, and generally minor (apples vs. pears) changes in tastes'.[5] What economists have neglected, according to Hirschman, is the 'non-wanton' kind of preference change, which 'is not really a change in tastes at all. A taste is almost defined as a preference about which you do not argue – *de gustibus non est disputandum*. A taste about which you argue, with others or *yourself*, ceases ipso facto being a taste – it turns into a *value*.'[6] And though Hirschman himself does not make the point, it is precisely the 'non-wanton' type of preference change that best seems to define the process of becoming modern in developing countries.

Before outlining how we intend to address the neglected (and related) aspects of the process of modernization that have just been described, it is appropriate first to set our concerns in the context of the many criticisms of traditional consumption theory that have already been made.

TRADITIONAL CONSUMPTION THEORY AND ITS CRITICS

In its usual textbook version, traditional consumption theory embodies a core set of well-known assumptions which are listed below: the consumer is assumed always to maximize utility (this occurs, formally, where the budget line is tangential to an indifference curve and requires that the marginal rate of substitution between any pair of goods be equated to the ratio of their prices); the consumer 'always prefers more of a commodity to less' – this is known as the condition of insatiability or non-satiation; the consumer has consistent and transitive preferences – 'Thus, if person A prefers the II batch of goods to the I batch, he does not at the same time prefer the I batch to the II. And if he prefers batch III to II to I, then he necessarily prefers III to I';[7] the consumer has given stable preferences; the consumer is sovereign – this requires that the preferences of the consumer not be shaped by any aspect of the process by which they are satisfied. The sovereignty of the consumer in this sense would be ensured by the assumption that his preferences are innately determined; the consumer's preferences can be inferred from his market behaviour (as postulated by revealed preference theory); the satisfaction that the consumer derives is solely a function of his own consumption, that is, he is in no way affected by the consumption behaviour of others; the consumer is fully aware of the range of commodities available and their prices, and also has complete knowledge of his own preference orderings.

Each of these assumptions has been the object of much criticism by economists and non-economists alike.[8] Some of the criticisms are concerned with the normative aspects of traditional theory, while others (arguably the majority) are concerned with the problems of the theory from the point of view of demand analysis.[9] Some of the critics concern themselves with methodological issues[10] while still others focus on the empirical failings of traditional theory.[11] A large number of these criticisms are based explicitly (in the case of the well-known critiques by Scitovsky, Hirsch and Galbraith[12]) or more commonly, implicitly, on the circumstances prevailing in the developed countries. A no less substantial body of literature has also questioned the relevance of conventional consumption theory (and neoclassical economics in general) in the very different context of highly traditional societies. For example, 'In reaction against what they have imagined to be the traditional, economists' view of "economic man", with his rational, calculating, maximizing behaviour, anthropologists have often tended to stress the extent to which economic behaviour in "primitive" societies has been governed by social criteria, moral rule, ritual proscription.'[13]

What have been much less frequently emphasized, however, are the limitations of traditional theory in coming to grips with *the transition from traditional to modern forms of consumption behaviour*. Part of this change process, as noted above and as Kuznets[14] has rightly emphasized, are continuous technical changes in products. The neglect of these changes (as well as the changes in values which may often be associated with them) is partly a reflection of the confining short-run focus of much of welfare economics. To quote Little, 'Over a comparatively short period, say a few years, most people's tastes will not change very much. Also the number of novelties introduced will not be great, and quality changes are not likely to be very important.'[15] But even restricting the scope of welfare economics in this way does not entirely eliminate the problem since technical changes in products *do* occur over 'a few years' with welfare consequences that need to be addressed. However, one has then (as in the longer run) to confront the fact that traditional consumption theory is fundamentally unsuited to deal with new products. In particular, 'Introduction of a new good requires that the preference function defined on *n* goods is thrown away and with it all the knowledge of behaviour based on it . . . or the fiction that the consumer has a potential preference function for all goods present and future . . . Neither approach gets us very far.'[16] Fortunately, the 'new' theory of demand, which treats products as bundles of characteristics offers a way out of this difficulty and the chapters below make extensive use of this approach.

Closely related to, but nevertheless distinct from, technical changes in products, are (as was also noted above) changes in values, or what Hirschman refers to as 'non-wanton' preference changes. Though these changes have been neglected by economists (at the expense he suggests of 'wanton' changes of preferences) they have been accorded much more attention by sociologists and social psychologists. Indeed, some of the writing in these fields treats values as a highly pivotal and integrative concept for the social sciences as a whole. Rokeach, for example, suggests that 'the value concept, more than any other, should occupy a central position across all the social sciences – sociology, anthropology, psychology, psychiatry, political science, education, economics and history. More than any other concept, it is an intervening variable that shows promise of being able to unify the apparently diverse interests of all the sciences concerned with human behaviour.'[17] At any rate, there is much that can already be learnt from those psychologists and sociologists who have studied how and why values change and especially from those who have examined these changes in the context of the modernization process in developing countries.

Most of the chapters below are concerned in one way or another with the technical changes in products or with the (related) changes in values that comprise essential features of the process of modernization. (Some chapters in fact deal specifically with the relationships between these two types of change.) The essays in Part One take as their point of departure the notion that as latecomers to development, today's developing countries are heavily influenced by technical changes in products that originate in the industrialized world (in much the same way and for many of the same reasons that developing countries are dependent on technical changes in processes or technologies from the rich countries). These chapters are concerned therefore not only with the various mechanisms through which technical changes in products are transmitted to developing countries, such as advertising (Chapter 4), product standards (Chapter 5), multinationals (Chapters 2, 3, 4, 6 and 7), international demonstration effects (Chapter 6) and colonialism (Chapters 5 and 6), but also with the welfare effects to which each such mechanism (or a combination of mechanisms) gives rise. These welfare effects may or may not involve changes in values and they refer to absolute as well as to relative gains and losses. In assessing these various welfare effects we contribute both methodologically and empirically to the debate that has often been joined over the effects of multinationals on developing countries,[18] since these enterprises are obviously an important mechanism through which consumption patterns are transmitted from rich to poor countries. (Chapter 3, for example, addresses the much debated question of whether and in what manner multinationals actually adapt products that originate in the developed countries to Third World circumstances.) However, as we are concerned at various points to emphasize, multinationals are by no means the only such transmission mechanism.

By relaxing various assumptions of traditional theory, our argument calls into question the welfare implication of this theory, that by extending the range of choice, technical change in products may improve the welfare of some consumers and at worst will leave other consumers unaffected. What we suggest in contrast is that although new products may indeed benefit some consumers (albeit in degrees that seem to favour the relatively affluent), there are a variety of ways in which these products (directly and indirectly) may lead instead to frustration and disappointment (See Chapter 2, and especially Chapter 7 which compares the various sources of frustration and disappointment with new products.)

All the essays in Part Two raise questions for policy that are ignored by (or misconceived according to) traditional consumption theory. A host of questions are raised, for example, by the recognition – derived from Part One – that countervailing measures are needed during the modernization process to redress the inegalitarian welfare impact of new products in developing countries. Some of these issues are discussed in Chapter 8 which focuses on measures to promote appropriate products.

Several of the chapters in Part Two address another policy issue that is highly relevant in the context of modernization, but which does not derive from the first part of the book. While that part deals largely with consumption in an urbanized, relatively modern setting, a major policy problem concerns the rural sector of developing countries in which traditional values and modes of behaviour are often highly resistant to change. Evidently, far from taking these values and modes of behaviour as given, policy-makers seek instead to change them over a wide range of areas such as health, nutrition and sanitation.[19] In these endeavours, policy-makers receive little help from either traditional consumption theory or the new household economics, concerned as they exclusively both are with 'minor' or 'wanton' preference changes. Indeed, according to Hirschman, one of the least desirable outcomes of this bias is that it has impeded 'a serious intellectual effort to understand what are strongly held values and difficult to achieve changes in values'.[20] By drawing on other disciplines, we suggest (in Chapters 9 and 10) a number of ways in which this bias might be redressed and how as a result more effective policies might be designed.

The last chapter argues that some of the most important aspects of the modernization process in developing countries – and in particular the relationships between products, technologies and incomes – can only be understood by a more disaggregated analysis than is allowed by traditional consumption theory.[21] On the basis of such an analysis Chapter 11 questions what has become the received view of the relationship between an income redistribution in favour of the poor and employment; in so doing, we call into question also the policy that might be implied by this view.

Notes

1. 'In economic analysis individual preferences seem to enter in two different roles: preferences come in as determinants of behaviour and they also come in as the basis of welfare judgements' (Sen, 1973), p. 253. As Graaff (1967) pointed out more than twenty years ago, welfare economics (unlike positive economics) has necessarily to be judged solely on the basis of the merit of its assumptions.
2. For a description of this process see, for example, Inkeles (1983).
3. See Kornai (1971), p. 138, who provides a list of some of the most important technical changes in products.
4. See Graaff (1967), Little (1960) and Winch (1971).
5. Hirschman (1986), p. 145.
6. Ibid. (emphasis in original). Hirschman uses this distinction to criticize the Stigler and Becker (1977) model, which assumes given and constant preferences. This model is criticized also by Blaug (1980), Cowen (1989) and Rosenberg (1985).
7. Mishan (1982), p. 43.
8. The utility maximization assumption has come under criticism from a number of psychologists who have studied the way in which individuals actually make decisions. Herrnstein and Mazur (1987), for example, argue that the constant rate of time discounting that this assumption requires is not empirically valid: they suggest a model in which time is discounted instead in terms of a hyperbolic function. Furthermore, they contend that consumers make decisions not on the basis of marginal utility but on the simpler and more practical basis of average utility. Tversky and Kahneman (1986) conclude, on the basis of a large number of experiments, that 'people exhibit patterns of preference which appear incompatible with expected utility theory' (p. 125).

   An important reason for this finding has to do with the framing of decision problems: 'changes of perspective often reverse the relative apparent size of objects and the relative desirability of options' (p. 123). See also Earl (1983) and March (1986).

   The non-satiability of individual preferences has been criticized by Leibenstein (1976) who makes use instead of the notion of consumption targets or standards, which, he believes, depend heavily on the 'social influence group' to which the household belongs.

The assumptions that preference orderings are consistent and stable are related, in so far as changes in preferences may lead to inconsistent decisions (that is, where good A is chosen with the preferences prevailing at time 1 and good B is chosen, given the same possibilities at time 2). Kornai (1971) has argued that a major cause of inconsistency is learning over time: 'The decision-makers attempt to lessen uncertainty by learning and collecting information. This results in frequent shifts in preference orderings over time' (p. 146). Learning is central also to the contributions made by Pollak (1978), Hirschman (1982), von Weiszacker (1971) and March (1986).

As noted above, the assumption of consumer sovereignty requires that the preferences not be shaped by any aspect of the process by which they are satisfied. It was also noted that exogeneity in this sense would be ensured by innately determined preferences. Cultural anthropologists, however, tend to view innate tendencies as only one determinant of human behaviour. That is, while such tendencies exist, they are moulded by a multiplicity of cultural factors, so that the observed behaviour of individuals is a joint (and non-additive) product of their innate endowments and their specific cultural circumstances. See, for example, Geertz (1973). Detailed criticisms of the notion of consumer sovereignty are to be found in Penz (1986), Hollis and Nell (1975), Galbraith (1973) and Scitovsky (1976).

Critics of the revealed preference assumption all reject in one way or another the notion of a single, all-purpose preference ordering which 'is supposed to reflect his interests, represent his welfare, summarize his idea of what should be done, and describe his actual choices and behavior' (Sen, 1977), p. 335. Sen, for example, emphasizes the concept of commitment, which he defines as individual choice that is expected to yield a lower level of welfare to the individual than an available alternative. He also emphasizes (1973) the form of social interdependence that is described as the 'prisoners' dilemma'. The problem here is that the market mechanism is entirely unable to give effect to the 'higher-order' or collective consumption preferences of the individual members of society. (See also Schelling (1978) on this issue.) Still another criticism is based on the recognition (Sen (1977), March (1986) and Schelling (1984)) that people are in fact made up of several 'selves' and that each of these facets of the same person may hold quite different preferences. For example, one self may often hold a 'metapreference' that is critical of the self who consumes unhealthy foods and smokes cigarettes. A number of authors have argued, moreover, that it is precisely the *unrevealed* preference that better represents the preferences of the consumer. The essence of this argument is that people are often prone to act 'against their own better judgment', as would occur for example when they suffer from a systematic 'temporal myopia' (Elster, 1986).

In rejecting the assumption of independent preferences, Hirsch (1976) and Frank (1985) argue that consumption takes a 'positional' form, in which gains for some are offset by losses to others. Because it can be formalized in terms of the 'prisoner's dilemma', positional consumption also leads to rejection of the revealed preference assumption.

The final assumption, of perfect information, means that there is never a discrepancy between the utility that the consumer expects to achieve and the utility that is actually realized. Among the most important concepts that this assumption neglects are 'acquired preferences' (Scitovsky, 1986) and 'disappointment' (Hirschman, 1982).

9. See Pollak (1978).
10. A fundamental methodological issue turns on the distinction between the concept of rationality employed by traditional economic theory on the one hand and by psychology and sociology on the other.

Traditional microeconomic theory employs what Herbert Simon refers to as 'substantive' rationality. 'Behaviour is substantively rational when it is appropriate to the achievement of given goals within the limits imposed by given conditions and constraints' (Simon, 1979), p. 67. Given the assumption of utility maximization and subject to the constraints of the consumer, the substantively rational solution follows as a matter of mere deductive logic. And, as Simon is at pains to emphasize, this concept of rationality has 'freed economics from any dependence on psychology' (Simon, 1979), p. 67.

In contrast, the concept of rationality that is said to be generally employed in psychology may be described as 'procedural'. As its name suggests, this form of rationality is concerned with the nature of the deliberative process that underlies behaviour in general and with consumer choice behaviour in particular. Where this process is sufficiently reasoned (or deliberative), behaviour is described as being rational. 'Conversely, behaviour tends to be described as "irrational" in psychology when it represents impulsive response to affective mechanisms without an adequate intervention of thought' (Simon, 1979), p. 68. Unlike the substantive rationality of traditional consumption theory, therefore, the psychologists' concept is concerned with the *process* rather than the *outcome* of choice decisions. And to understand the former, psychologists find it necessary to examine the numerous factors that determine how consumption decisions are *actually* made.

Scitovsky's critique of traditional micro theory is in fact based on a large body of evidence from behavioural psychology; he focuses on the process whereby the individual seeks to raise his level of arousal from a level that is too low, because of inadequate stimulation (Scitovsky, 1976).

The concept of rationality employed in sociology is no less firmly grounded in the empirical study of decision-making. But as Smelser points out the sociological concept is 'something more than a psychological postulate; it is a standard of behavior to which people conform or from which they deviate . . . Viewed in this way, economic rationality becomes a type of social control . . . Furthermore, when viewed in this way, the notion of economic rationality becomes an empirical variable rather than a fixed postulate, from the *methodological* point of view' (Smelser, 1976, pp. 3–4, emphasis in original).

For an interesting comparison between the way economists view preferences and the approaches taken by other disciplines, see Etzioni

(1985). This article also draws out some of the normative implications of these differences. See in addition the readings in Huppes (1976) and the textbook on economic psychology by Lea, Tarpy and Webley (1987).

11. See, for example, Abramovitz (1979).
12. See Scitovsky (1976), Hirsch (1976) and Galbraith (1973).
13. Firth (1967), p. 8. Also see Douglas and Isherwood (1979) and Sahlins (1972).
14. Kuznets (1967).
15. Little (1960), p. 48.
16. Lancaster (1966), p. 20.
17. Rokeach (1973), p. 3. See also Rokeach (1979) and Feather (1975).
18. Jenkins (1988).
19. See Johnston and Clark (1982) and Rogers (1983).
20. Hirschman (1986), p. 145.
21. See Stewart (1977), James (1976) and the appendix to Chapter 11.

References

Abramovitz, M. (1979). 'Economic Growth and its Discontents', in Boskin, M. (ed.), *Economics and Human Welfare*, Academic Press.

Blaug, M. (1980). *The Methodology of Economics*, Cambridge University Press.

Cowen, T. (1989), 'Are All Tastes Constant and Identical?: A Critique of Stigler and Becker', *Journal of Economic Behavior and Organization*, Vol. 11.

Douglas, M. and Isherwood, B. (1979). *The World of Goods*, Basic Books.

Earl, P. (1983). *The Economic Imagination*, Wheatsheaf.

Elster, J. (ed.) (1986). 'Introduction' to *Rational Choice*, New York University Press.

Etzioni, A. (1985). 'Opening the Preferences: A Socio-Economic Research Agenda', *Journal of Behavioral Economics*, Winter.

Feather, N. (1975). *Values in Education and Society*, Free Press.

Firth, R. (ed.) (1967). 'Themes in Economic Anthropology: A General Comment', in *Themes in Economic Anthropology*, Tavistock.

Frank, R. (1985). 'The Demand for Unobservable and other Nonpositional Goods', *American Economic Review*, March.

Galbraith, J.K. (1973). *The Affluent Society*, Pelican.

Geertz, C. (1973). *The Interpretation of Cultures*, Basic Books.

Graaff, J. de V. (1967). *Theoretical Welfare Economics*, Cambridge University Press.

Herrnstein, R. and Mazur, J. (1987). 'Making Up Our Minds: A New Model of Economic Behavior', *The Sciences*, Nov./Dec.

Hirsch, F. (1976). *The Social Limits to Growth*, Harvard University Press.

Hirschman, A. (1982). *Shifting Involvements: Private Interest and Public Action*, Princeton University Press.

Hirschman, A. (1986). *Rival Views of Market Society and Other Recent Essays*, Viking.

Hollis, M. and Nell, E. (1975). *Rational Economic Man: A Philosophical Critique of Neoclassical Economics*, Cambridge University Press.
Huppes, T. (ed.) (1976). *Economics and Sociology: Towards an Integration*, Martinus Nijhoff.
Inkeles, A. (1983). *Exploring Individual Modernity*, Columbia University Press.
James, J. (1976). 'Products, Processes and Incomes: Cotton Clothing in India', *World Development*, Vol. 4 (2).
Jenkins, R. (1988). 'Transnational Corporations and Third World Consumption', *World Development*, Vol. 16 (11).
Johnston, B. and Clark, W. (1982). *Redesigning Rural Development*, Johns Hopkins.
Kornai, J. (1971). *Anti-Equilibrium*, North-Holland.
Kuznets, S. (1967). *Modern Economic Growth*, Yale University Press.
Lancaster, K. (1966). 'Change and Innovation in the Technology of Consumption', *American Economic Review*, May.
Lea, S., Tarpey, R. and Webley, P. (1987). *The Individual in the Economy: A Textbook of Economic Psychology*, Cambridge University Press.
Leibenstein, H. (1976). *Beyond Economic Man*, Harvard University Press.
Little, I.M.D. (1960). *A Critique of Welfare Economics*, 2nd edn, Oxford University Press.
March, J. (1986). 'Bounded Rationality: Ambiguity, and the Engineering of Choice', in Elster, J. (ed.) *Rational Choice*, New York University Press.
Mishan, E. (1982). *What Political Economy is All About*, Cambridge University Press.
Penz, P. (1986). *Consumer Sovereignty and Human Interests*, Cambridge University Press.
Pollak, R. (1978). 'Endogenous Tastes in Demand and Welfare Analysis', *American Economic Review*, Vol. 68 (2).
Rogers, E. (1983). *The Diffusion of Innovations*, 3rd edn, Free Press.
Rokeach, M. (1973). *The Nature of Human Values*, Free Press.
Rokeach, M. (1979). *Understanding Human Values*, Free Press.
Rosenberg, A. (1985). 'Prospects for the Elimination of Tastes from Economics and Ethics', in Paul, E.F. *et al.* (eds) *Ethics and Economics*, Blackwell.
Sahlins, M. (1972). *Stone Age Economics*, Aldine.
Schelling, T. (1978). *Micromotives and Macrobehavior*, W.W. Norton.
Schelling, T. (1984). *Choice and Consequence*, Harvard University Press.
Scitovsky, T. (1976). *The Joyless Economy*, Oxford University Press.
Scitovsky, T. (1986). *Human Desire and Economic Satisfaction*, New York University Press.
Sen, A. (1973). 'Behaviour and the Concept of Preference', *Economica*, Vol. 40, August.
Sen, A. (1977). 'Rational Fools: A Critique of the Behavioral Foundations of Economic Theory', *Philosophy and Public Affairs*, Vol. 6.
Simon, H. (1979). 'From Substantive to Procedural Rationality', in Hahn, F. and Hollis, M. (eds), *Philosophy and Economic Theory*, Oxford University Press.
Smelser, N. (1976). 'On the Relevance of Economic Sociology for Econ-

omics', in Huppes, T. (ed.), *Economics and Sociology: Towards an Integration*, Martinus Nijhoff.

Stewart, F. (1977). *Technology and Underdevelopment*, Macmillan.

Stigler, G. and Becker, G. (1977). 'De Gustibus Non Est Disputandum', *American Economic Review*, March.

Tversky, A. and Kahneman, D. (1986). 'The Framing of Decisions and the Psychology of Choice', in Elster, J. (ed.), *Rational Choice*, New York University Press.

Weiszacker, C. von (1971). 'Notes on Endogenous Technical Change', *Journal of Economic Theory*, Vol. 3.

Winch, D. (1971). *Analytical Welfare Economics*, Penguin.

Part One
The Influence of the Developed Countries on Consumption and Welfare in the Third World

2 New Products: A Discussion of the Welfare Effects of the Introduction of New Products in Developing Countries*

With Frances Stewart

INTRODUCTION

One of the most important ways in which advanced countries have changed poor countries is through the impact of new products. Products designed in industrialized countries have transformed consumption and production patterns, culture and society. The significance of new products is widely recognized at a popular level and forms a central part of much informal discussion: 'good' products (vaccination, new seeds) are contrasted with the notoriously 'bad' (powdered baby milk, high-tar cigarettes, sophisticated and expensive weapons). But while the effects of new products are recognized as of enormous importance at a popular level, economists have generally neglected the question. Discussions of technology transfer are mostly framed in terms of the costs of the transfer and of the production characteristics of the techniques.

The neglect of the effect of new products in LDCs in large part follows a similar neglect in advanced countries. Tens of thousands of new products are produced and marketed each year and as many more 'old' products are changed in significant respects.[1] Indeed, the very dynamic of oligopolistic competition requires adherence by firms to the maxim of 'striving to render obsolete everything you have done before'.[2] Yet most discussion of economic welfare ignores new products, or changing characteristics of old products.[3] For example, in neither of two classic works on welfare economics, Little or Graaff, is there any substantive discussion of new products. In so far as new products are discussed in traditional consumer theory, the most that

* We are grateful for comments on an earlier draft from Keith Griffin, Paul Isenman, Peter O'Brien and Howard Pack. Ruchira Chatterji provided valuable research assistance.

is said is that – by increasing choice – consumers' welfare is potentially increased. A significant reason for this lacuna is that the traditional theoretical framework is unsuited, perhaps incapable, of dealing with the question of new products.[4] But the 'new' theory of consumer demand, associated with Lancaster, which treats products as bundles of characteristics, lends itself far more readily to dealing with the question.[5]

The aim of this chapter is to suggest a more systematic framework in which to assess the impact of new products on poor countries. To do so we consider the general conditions according to which new products are likely to increase/decrease economic welfare, looking both at 'absolute' effects and also distributional consequences. We consider the likely direction of effects when products from advanced countries are introduced into poor countries. We briefly describe and analyze eight case studies. Finally we consider some of the policy implications of our findings.

THE LANCASTER FRAMEWORK AND NEW PRODUCTS

Each product is regarded as being a bundle of characteristics, combined in a certain ratio, so that each product may be represented as some vector **q**, representing a certain combination and quantity of characteristics, i.e. **q** = (i*a*, ii*b*, iii*c* . . .) where *a*, *b*, *c*, describe the characteristic of the products (such as nourishment, beauty, etc.) and i, ii, iii, the quantity of each characteristic embodied in the product. It is assumed that consumer preferences are ordered for *characteristics* rather than for products in themselves. A new product – represented say by some vector, **r** – then consists in a changed bundle of characteristics. Since the consumer ordering is for characteristics, not products, the new bundle of characteristics may be compared with the old, without requiring a new consumer ordering. In contrast, in traditional theory, since consumer preferences are assumed to be for products not characteristics, a new product must involve a new ordering. Since a reordering leads to non-comparability between the two situations being considered it rules out welfare consideration of new products. This is the fundamental reason why traditional theory cannot deal with new products, but the new theory can. It should be noted that the new approach does not distinguish between changed characteristics of old products and 'new' products – both would be represented by a different vector involving a changed bundle of characteristics. This lack of distinction seems reasonable in the light of the gradual spectrum of changes which may occur, so that any classification of changed products into 'old' products with changed characteristics and 'new' products is somewhat arbitrary. In what follows, therefore, we phrase the discussion in terms of the impact of 'new' products, although the discussion also encompasses what might more normally be described as minor changes in old products. This means that the discussion has a much wider application – empirically – than a discussion of new products confined to major product innovations, such as the motor car, the ball point pen or the microwave oven. Indeed it could be said that in modern societies, almost all products change in some way in a quite short time – the exceptional products are the ones that *do not* change.

While, as we show in some detail below, the framework is much more amenable to handling the question of new products, there remain problems of non-comparability when new products involve entirely *new* characteristics. Of course if characteristics are very broadly defined – e.g. nourishment, warmth, entertainment, etc. – it might be possible to avoid this problem, since at the broadest level human needs – basic and base – and therefore product characteristics, remain similar over time. But with such broad definitions it may be difficult to pick up various subtle changes in product characteristics which are of significance: 'washing whiter', or 'speeding up calculations' *can* be classified under very broad categories such as 'satisfaction in appearance' or 'efficiency of operations', but something is lost in so doing. Thus where a new product involves a new characteristic we may be back to the problem of non-comparability, associated with the old approach. But this is likely to happen in a much smaller range of cases, and may be largely avoided by a broad classification of characteristics. In what follows we ignore this problem. For simplicity of presentation, we consider products with just two characteristics.

With the new approach problems of comparison arise if the introduction of a new product directly, or as the result of associated promotional efforts, alters tastes and hence consumer ordering. To start with we assume that changes in consumers' welfare are solely a function of an individual's own consumption, and that tastes are unchanged as a result of the introduction of new products. In such a situation, a new product will unambiguously improve economic welfare if it extends the range of efficient choices by creating a new bundle of characteristics, containing more of at least one characteristic for a given cost while leaving the existing range of products unaffected, in both availability and cost.

Figure 2.1 portrays a situation of two characteristics, 1 and 2, and

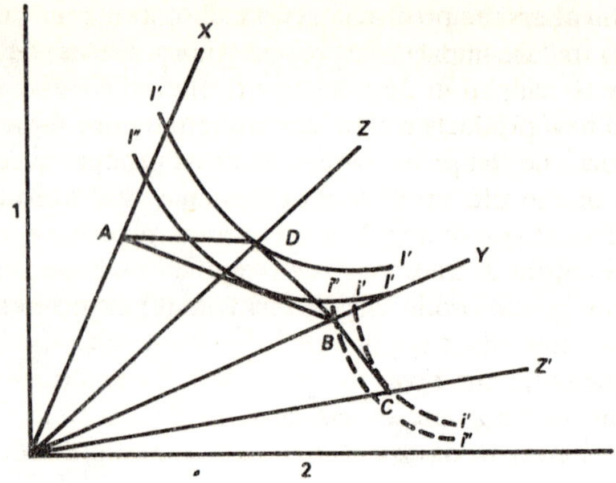

Figure 2.1

two initial products, *X* and *Y*. *AB* represents the price ratio between the two goods. New product *Z* represents a combination of characteristics previously obtainable by combined consumption of *X* and *Y*. The price of *Z* is such as to extend the efficiency frontier from *AB* to *ADB*. *Z'* represents another new good which embodies a previously unobtainable combination of characteristics and thus extends the range of efficient choice. Since neither *X* nor *Y* is displaced, the introduction of either (or both) of these new goods represents a Pareto improvement irrespective of the preference pattern of consumers. However, the preference pattern of consumers will determine the *distribution* of gains from the new products. With the preference pattern shown in the diagram as *I'I'* and *I''I''*, consumers gain by the new good *Z*, but are not affected by *Z'*. However, consumers with preference pattern *i'i'* and *i''i''* would gain by *Z'* but are not affected by *Z*.

In many situations, the introduction of a new product either totally displaces existing products, or leads to altered production and cost conditions for them. If the new product involves the same ratio of characteristics as the old product, but a greater quantity for any given cost, then its introduction is likely to lead to the displacement of the old product as 'rational' consumers will no longer consume it: since consumers will be able to consume more of both characteristics for the same cost, in a Pareto sense welfare will have improved.

But if a new product involves a *changed* combination of characteristics for a given cost (more of some characteristic and less of another), *and* alters the production cost and/or availability of existing products, its welfare impact is more ambiguous. Generally speaking, this is likely to happen in three types of condition. First, where the existing and new products compete in using the same (scarce) factor of production, and the production of the new product raises costs of production of the old product. For example, the introduction of mechanized methods of textile manufacture may have raised the price of raw cotton to traditional spinners. At a more general level, the vast array of new products (and technology) introduced over the past 200 years has raised the price of labour to producers of existing products, increasing their costs of production. Secondly, where there are economies of scale, and diseconomies of small scale, so that if the demand shifts away from existing products to new products, costs of production of existing products rise.[6] Thirdly, where markets are dominated by a few very large firms, these firms may withdraw their old products (and/or raise their price) when they introduce new products. They may do so because costs have risen, for one or other of the reasons just mentioned, or because their marketing strategy requires it. The latter is probably the most usual reason with firms withdrawing old models so as to ensure that demand is switched to the new products. Changes in fashion with respect to clothing and automobiles are obvious examples here.[7]

If, for any of these reasons – competing use of scarce resources, indivisibilities or marketing strategy – the relative costs of old products rise, or they become non-available, then the introduction of the new products may lead to a deterioration in welfare, for some groups in society, or even for whole societies.

In Figure 2.2, the initial efficiency frontier was represented by AB. The introduction of a new product, Z, leads to a rise in the cost of production of X, such that the new frontier is represented by $A'CB$. Whether consumers are made better or worse off by the change will depend on whether their indifference map is represented by the set of II, or by ii. Consumers with indifference curves ii will be better off; those with the II map worse off. Any situation in which the price of existing products is increased, or their production withdrawn, may make some consumers worse off, if the new products involve a changed bundle of characteristics (for a given cost).

If all consumers had identical indifference maps, then they would either *all* be of a II type, or all of a ii type. If they were all of a ii type then they would all be better off. If all were of a II map, then presumably the new product would not be introduced. Hence, for

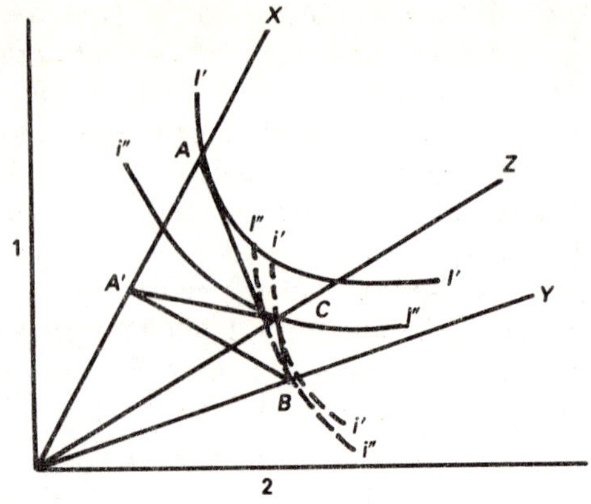

Figure 2.2

new products to involve a deterioration in welfare, consumer preferences must differ, and while the welfare of some consumers may go down, that of others goes up. As one would expect, the group of consumers whose preferences are most directed towards the characteristics combined in the old product, whose price has risen (or which has been withdrawn), are most likely to suffer a deterioration in welfare, while those whose preferences are most directed at characteristics embodied in the new goods are most likely to gain. In a commonsense way we can say something about the likely causes for people to be in one group or the other:

1. People with 'conservative tastes' who dislike change are likely to be among those with preferences weighted towards the old product characteristics, and are therefore more likely to be losers.
2. In so far as economies of scale are of significance in determining the change in costs, then those with the tastes of the majority are likely to be among those who gain, those with minority tastes among those who lose. One should add that 'minority'/'majority' is here defined with reference to total effective *monetary* demand, not with respect to felt needs. With an unequal income distribution, the majority of the people may represent the minority of the purchasing power.
3. One of the most significant sources of differences in tastes is that of differing income levels. As consumers get richer, the product characteristics of the goods they consume change towards more high-income types – i.e. better quality, more uniform standards,

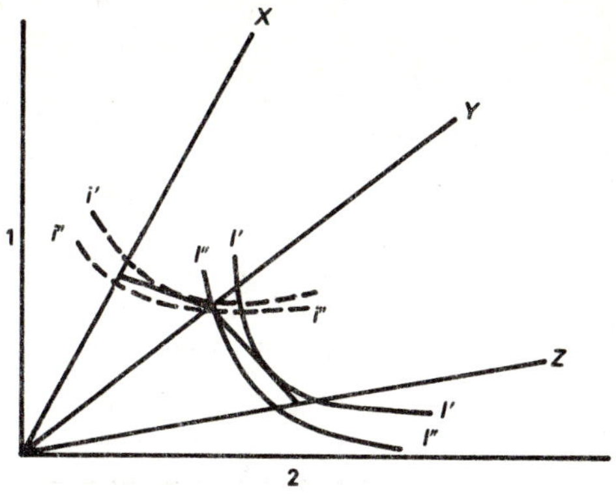

Figure 2.3

more luxurious etc. (see Stewart, 1977). In so far as new products are directed to satisfy the needs of those with rising incomes, then each new set of products will embody more high-income characteristics and less low-income characteristics. Hence the higher-income consumers will tend to be among those who gain, the low income among those who lose.[8]

Suppose characteristic 1 represents low-income characteristics, characteristic 2, high income. Then a new product, *Z*, is likely to be to the south-east of *Y* representing more high-income characteristics. Where the new product leaves existing products unaffected, no group of consumers will lose, but the gains from the new product will be concentrated among high-income consumers. Suppose high-income consumers have indifference map *II* in Figure 2.3. With the introduction of new product *Z*, they will move from *I″I″* to the higher indifference curve *I′I′*. But the poor consumers with indifference map *ii* will remain on the same indifference curve, *i′i′*. Thus while no one will lose, the relative gains will be concentrated on the rich consumers. If as a result of the new product, one of the old products, *X*, is withdrawn from production, then poor consumers (indifference map *ii*) will suffer a loss of welfare moving to a lower indifference curve *i″i″*. A similar consequence would follow for an increase in cost of the old products. If the new product represented a more *efficient* product (i.e. involving at least as many of each characteristic as some existing product, at lower cost) the old product would be eliminated as inefficient. *All* groups of consumers would gain. However, unless each characteristic was reduced in cost by the same proportion,

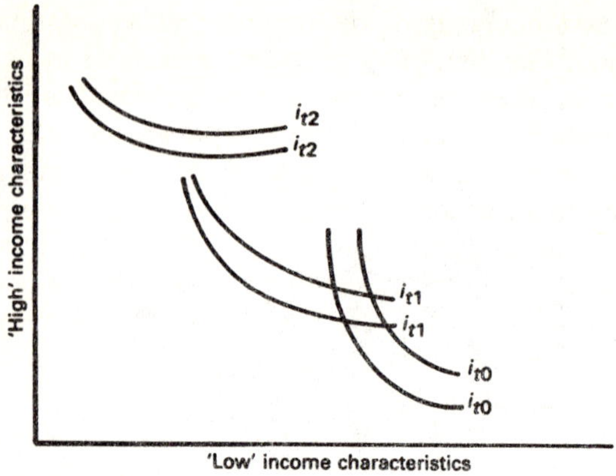

Figure 2.4

relative gains would be concentrated among the consumers who most prefer the characteristic most reduced in cost.

Over time, as incomes rise there is a systematic shift in consumer preferences towards higher-income characteristics, so that 'average' preferences shift as shown in Figure 2.4.

New products are designed – and old products withdrawn – in accordance with this shift in tastes. When new products are introduced into an economy for which they are designed (i.e. when technological change is endogenous) negative effects on consumers' welfare, of the type described above, are likely to be confined to certain groups within society, because if *all* consumers were made worse off, then it would be unlikely that the new products would be developed and introduced. This is generally the situation in advanced countries. Minority groups and poorer consumers may suffer welfare deterioration as new products are introduced but majority (in a monetary demand sense) consumers will gain: an obvious example is the way in which poorer consumers lose as private transport replaces public, and large-scale units of consumption replace small. To assess the overall welfare impact of new products in such cases requires weighting the gains and losses of different groups of consumers.

The situation is somewhat different when products from one (richer) society are introduced to another (poorer) society. New 'high-income' products are developed for the high- and rising-income societies, and consequently have increasingly high-income characteristics. While these new products tend to correspond to the consumer preferences of large groups in the rich societies – i.e. in terms of

Figure 2.3 the *ii* map may represent all (or most) consumers in poor societies; the *II* map that of all (or most) consumers in rich societies. If that were so, then the question to be answered is why the new products should be introduced into the poorer societies at all, and why, if they are (e.g. to serve a minority elite with similar consumption patterns to those of the rich society), the low-income goods should be increased in cost or withdrawn. There are of course goods where this does not happen, and countries where the new products are not introduced. And in most poor countries there is some lag before new goods are introduced from rich societies. But there is some tendency for the new high-income goods to be introduced and old goods to be withdrawn even where the new goods do not correspond to the main consumer demand. This may happen:

1. Where monetary demand in the poor country is heavily concentrated among the rich, so that preferences expressed by monetary demand are similar to those of the rich countries. Some Latin American countries provide well-known examples: it has been argued, by Furtado for example, that the maldistribution of monetary income and demand is a deliberate policy-induced strategy in order to create demand for the new products and therefore a market for those companies producing the new products. In this type of situation the negative welfare effects of new products arise from the distribution of income, rather than the introduction of new products as such (although the production technology associated with the new products – see below – may be partly responsible for the skewed income distribution).
2. Where production of *all* goods – old and new – takes place in the advanced countries: the rise in price of the old goods/withdrawal of them that arises with switches to the new goods in the advanced countries then also affects consumption possibilities in the poor countries.
3. Where production of both types of goods (old and new) takes place in the poor countries, using technology from the advanced countries, and the change in the advanced countries towards the new products changes the relative costs of production *in the poor country*: this could happen where products compete for scarce factors, or where there are world-wide economies of scale (e.g. in technology development, maintenance, inputs, advertising and so on), which affect the relative costs of production in the poor countries, even though the change in demand and production

initially only takes place in the rich country. In many cases, for example, problems (including rising costs and non-availability) in securing parts and inputs persuade entrepreneurs in poor countries to switch to more recent techniques and products.

4. Where the production of both goods, which takes place in the poor country, is dominated by particular decision-makers whose private interest lies in switching to more recent products. For example, a continuous switch in products may be the most effective way for multinational companies to prevent local competition.
5. Where production of both goods takes place in the poor countries – or where production of the old good takes place in the poor country, and the new good in the rich country – and where some switch in demand away from the existing products raises their costs – because of indivisibilities. Such a situation (ignoring for the moment the impact of advertising) may occur so long as there are some high-income consumers whose tastes are more in line with rich-country tastes (i.e. a similar situation to (1) above, but less extreme).

In two of these cases (1) and (5), some distributional judgments are involved in assessing the impact of new products, since some consumers gain: however, in both the distributional judgment required is of a different kind from that involved in the rich-country case. In case (1) in monetary terms, the majority of consumers gain, but the majority of consumers, in terms of *numbers* of people, may lose. This is unlikely to be the case in the rich-country situation. In case (5), the gainers may be confined to a minority among monetary consumers. In the rich-country situation, it is likely that new products will only be introduced if a substantial proportion of consumers are likely to gain. In the other cases ((2)–(4)), new products may be introduced even though the *whole* society loses. It should be noted that this will only be the case on certain assumptions about consumer preferences, about the nature of new products, and of indivisibilities associated with production. Whether or not these assumptions apply will vary in different products/countries. But it is possible to generalize a little:

1. Assuming consumer preferences are unaffected by the existence of alternative products and by promotional efforts (an assumption we relax below) poor-country preferences are likely to be towards low-income characteristics compared with rich-country products.
2. If new products are first developed *for* rich societies – as for the

most part they are – each new product is likely to embody more high-income and less low-income characteristics than existing products. However, because of technological advances in product design and technology, the cost per unit of each characteristic (including low-income characteristics) *may* be lower than that of old products.

3. There is abundant evidence of economies of scale in much of modern technology – which is accentuated by even more marked economies of scale in marketing etc.[9] In so far as one 'modern' technology product replaces another – as in much of the changes in the modern sector – indivisibilities may be of significance. Indivisibilities are much less marked in traditional techniques/products, so that costs of production of these traditional products may not rise as new modern-sector competing products are introduced. But costs of traditional products may rise if the costs of inputs rise as a result of the new products. The established decline in traditional products over time may in part be due to the fact that they do not represent efficient alternatives; in part to policy changes which discriminate against them (e.g. standard setting, making low brewing illegal, etc.). But part is likely to be due to rising costs. Traditional/old products may be more likely to survive in large economies, where even declining demand may still represent a sizeable market, than in small.

Taking these considerations together, whether or not new products benefit or harm poor societies is likely to vary according to the products. This is part of the explanation of why there are apparently 'good' and 'bad' products. But there are three other types of reasons, not allowed for in our framework so far: the effects of new products on tastes, learning by consuming effects, and indivisibilities and complementarities in consumption.

TASTE CREATION

Producer sovereignty type effects, associated with promotional efforts on the part of producers, mean that producers attempt (and often succeed) to change preferences. These attempts are systematically related to new products, so that expenditure to promote new products tends to increase preferences for new products as against old. This is particularly marked where the old products are traditional ones, normally associated with minimal promotion, but is also true where one new modern-sector product replaces another. One effect of the tendency for greater promotion of new products will be to accelerate the switch in demand away from old products and therefore to increase the tendency for these to rise in cost/be withdrawn. In a strict consumer-preference framework, it is not possible to compare a situation in which consumer preferences have been altered, with the preceding situation. But stepping outside this framework, it is reasonable to argue that promotional expenditure on one type of product is likely to lead to over-consumption of that product compared with the less-well promoted products. Consumers subject to such imbalanced promotional efforts will tend to spend a greater amount on the new products than they would in the absence of such expenditure. If we take expenditure patterns in the absence of promotion (or with 'equal' promotion, which may not produce the same results) as the 'balanced' pattern of consumption, then promotional expenditure on new products will tend to lead to an imbalance of consumption expenditure towards new products – which in the case of poor countries is likely to mean towards high-income products. Consumers will then spend too little on low-income characteristic goods. The results are likely to be insufficient expenditure on basic needs. It is plausible to argue that this has been the effect in relation to many goods consumed in poor countries.

It should be noted that the effect on preferences is not just a matter of formal promotional expenditure: all sorts of other influences also operate. For example, cultural dependency and the image of the sophisticated Westerner (which used to be described as 'demonstration effects') may have as much influence in promoting new products, as formal advertisements.

LEARNING BY CONSUMING

The formal theory of consumer preferences does not allow for learning by consuming. In practice this is a very significant element in consumption practices. Consumers become more efficient *qua* consumers as they consume – learning how the items consumed endure, how to use them, their effects in terms of the fulfilment of the needs for which they were demanded. These learning effects are a function of experience in consuming particular products: as a new product is introduced consumers tend to be inefficient in choice and in use once selected (i.e. secure less than the maximum possible combination of desired characteristics.) As time proceeds learning effects take place, and consumption efficiency rises. Since learning effects are related to the time for which a product has been available, the gap between efficient and actual consumption patterns is greatest for the most recently introduced products, least for the oldest products. In addition to this, there is a more general society-learning process, so that some societies are more efficient consumers than others, particularly in relation to certain types of products. In all societies, therefore, one can expect more efficiency in consumption of old products than new, and the gap in efficiency will be greatest where the difference in characteristics of products is greatest, and where the old products have been consumed for the longest period. One would therefore expect the biggest gap in efficiency between consumption of traditional products – which have been available for a long time and for which consumption efficiency is high – and modern newly introduced products. The observed fall in nutritional standards following the introduction of 'modern' processed goods is partly due to lack of learning experience with this type of product, as against the traditional products. The baby-milk case is an example of this. Rich societies in general put considerable effort into improving consumption efficiency – as a response to the steady introduction of new products, for which learning experience would otherwise be low. Children are educated to be discriminating consumers, consumer groups establish magazines etc. to inform consumers, and the government intervenes to protect and inform consumers. The net result is a much higher degree of consumption efficiency in relation to new products than in poor societies. Lacking the learning experience related to many new products, social (including government) protection and information services, there is a systematic tendency for inefficiency of consumption of new products in poor countries.

The combination of imbalanced promotional efforts and consumption inefficiency both concentrated on new products would tend to result, simultaneously, in an undervaluation of old products and an overvaluation of new products in relation to their true characteristics. This is illustrated in Figure 2.5. X and Y are existing products: Z is the new product. Because of differential promotional expenditure consumers underestimate the traditional products: whereas the actual combination of characteristics possible with a certain income is represented by ABC, and therefore the best position is on $I'I'$, consuming only X and Y, the potential of the traditional products

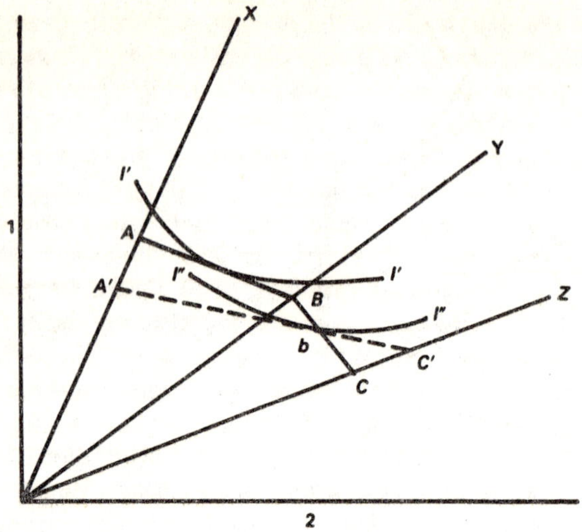

Figure 2.5

is underestimated, because of lack of information/advertising, while the potential of new product *Z* is exaggerated because of promotional efforts plus lack of learning effects. Consumers believe the efficiency frontier is represented by *A′C′* and therefore consume at *b*, on a lower indifference curve than they might have done in the absence of new product *Z*, or with correct information about the alternatives.

INDIVISIBILITIES IN CONSUMPTION

Many products are more or less indivisible, the extent of indivisibility varying with the product. At one extreme, food grains may be perfectly divisible (although not the packages in which they are purchasable); at the other a modern aircraft or spaceship can only be purchased in very large units. The higher the income of consumers, the larger the unit in which they are prepared, and indeed may wish to buy characteristics. This fact is reflected in the design of products – which tend to be increasingly indivisible[10] – in relation to any desired set of characteristics – the higher the income for which the products are designed. Indivisibility is thus a high-income characteristic. But it is sufficiently important to merit separate treatment. Products designed for high-income consumers tend to be more indivisible than would be appropriate for low-income societies. This means that new

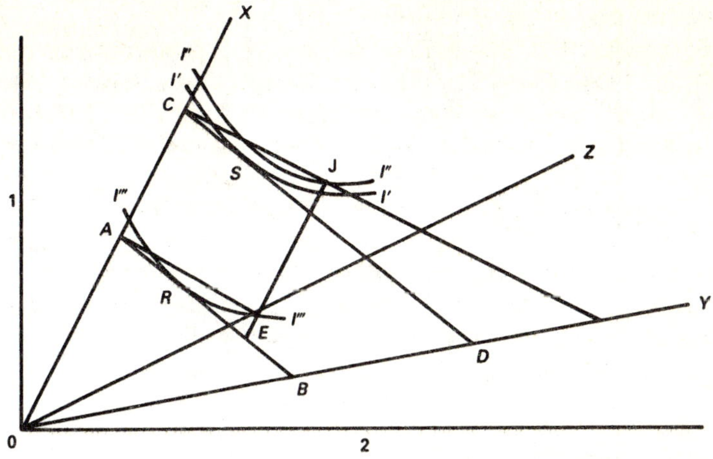

Figure 2.6

products will tend to benefit rich individuals more than poor.

Figure 2.6 shows two individuals with different incomes (represented by *AB* and *CD* respectively) and identical tastes (represented by the indifference map *II*). Prior to the introduction of new good *Z* the two individuals consume at points *R* and *S*. Good *Z* is now introduced but cannot be bought in units less than *OE*. The result is that the low-income individual cannot improve his welfare (i.e. he can only move to *E* on indifference curve I‴I‴) while the second individual is able to combine *OE* of good *Z* with *AC* (= *EJ*) of good *X* to reach *J* on the higher indifference curve *I″I″*.

If the indivisible item is none the less purchased by the poorer consumer because he wants certain of the associated characteristics, then this involves disproportionate expenditure on that item as against other items of consumption. Consumption patterns in some underdeveloped countries illustrate the effects of indivisibilities. Consumer durables are outstanding examples of indivisible consumption items. Wells' survey of expenditure on consumer durables in Brazil revealed very widespread ownership of consumer durables, even among relatively low-income groups.[11] Increasing consumption of consumer durables was accompanied by a decline in nutritional standards. A survey in São Paulo found that between 1959 and 1969/70 'there was a significant rise in the share of household expenditure devoted to domestic appliances . . . an analysis of (physical) *per capita* food consumption shows that there was a downward trend, indicating that levels of nutrition among the working class deteriorated absolutely'.[12] Analysis of two family expenditure surveys in 1968 and 1974 showed that the share of expenditure devoted to food fell from 46.7 to 41.1 per cent between 1961/62 and 1971/72. This shift in expenditure was accompanied by an absolute deterioration in nutrition standards, with 2704 calories being consumed in 1961/62 (73.7 g proteins, 75.6 g fats) compared with 2531 calories in 1971/72 (80.2 g proteins, 70.5 g fats).[13] These trends in consumption expenditure are of course the outcome of a variety of effects, besides indivisibilities. Others include changing behaviour in response to the availability of new products, to promotional expenditure and to demonstration effects.

COMPLEMENTARITIES OF CONSUMPTION

For many consumer goods there are interdependencies, such that the enjoyment of one good depends on the consumption of some other good: tennis balls and tennis rackets; cars and roads; chairs and tables; and so on. Interdependencies in consumption may reinforce some of the other effects already discussed because they mean that a new product and/or the withdrawal of an old product not only has direct effects on consumers, but also affects the welfare gained from other items. Thus some existing products may be rendered obsolete by the replacement of a complementary new product. The initial replacement of one good by another might apparently benefit all consumers, leading to more of all characteristics than the replaced product; but the complementarity effect may render obsolete other products which are replaced by goods which make some consumers worse off. (Of course the reverse may also happen.) Alternatively, complementarities may, like indivisibilities in consumption, cause the benefits from new products to be out of reach of low-income consumers, who cannot afford all items in the package. Complementarities may also reinforce the effects of new products in causing imbalance in consumption. Absence of complementary goods and services is a major factor behind the harm done by powdered baby milk. The practice of feeding powdered milk to babies 'is being used by women with no knowedge of hygiene, no ability to read the instructions on the can, and no money with which to buy sufficient powdered milk. Thus, diluted powdered milk from dirty bottles and dirty teats is substituted for breast milk. This leads to malnutrition and dietary disorders such as marasmus, diarrhoea and vomiting'.[14]

Cars and roads provide another example of indivisibilities cum complementarities. Modern cars generally (despite the success of the East African Safari) require modern roads for efficient use. If the cars are provided without the roads, then serious inefficiencies result. If the roads are also provided, social expenditure is concentrated on the provision of roads to car-owners, indirectly depriving other consumers of much needed infrastructure.

SUMMARY OF EFFECTS

The analysis suggests that the impact of new products on economic welfare will depend on the particular circumstances and will need to be assessed separately for each product and for each country – since the effects vary according to consumer tastes and income levels, and the product characteristics and production technology of new and existing products. In general we may distinguish the following situations.

A Where new products are likely to increase welfare in a Pareto sense. This occurs in two sets of circumstances:
- A_1 where the new products are more efficient than the old – i.e. each characteristic is of lower cost;
- A_2 where the new product does not displace nor add to the costs of existing products, and 'tastes' are not affected.

B where new products are likely to decrease welfare for some or all consumers. This may occur:
- B_1 where a new product displaces an old product or increases its relative cost;
- B_2 where promotional expenditure on the new product leads to over consumption.

In all of these cases the new products will have distributional consequences. In general where new products are designed in rich societies and introduced into poorer societies, the distributional consequences can be expected to be inegalitarian. Gains will be relatively concentrated among high-income consumers; losses relatively concentrated among low-income consumers. Thus we may distinguish two alternative distributional consequences, C.

C_1 where the new product is egalitarian in its consumption effects;
C_2 where the new product is inegalitarian in its consumption effects.

In general, data for changes in income distribution over time do not allow for these effects, since they look at shares of incomes received by different percentiles, not at the purchasing power in terms of characteristics. Even estimates (which are rare), which take into account changes in prices of the main product groups consumed by different income groups, do not disaggregate sufficiently to allow for the effects of the introduction of new products.

EFFECTS ON PRODUCTION TECHNOLOGY

The discussion above has been confined to considering the effects of new products on consumers *qua* consumers. In practice, an equally – perhaps more – significant effect on welfare will result from the effects on production technology, and hence on employment and income distribution. New products developed in advanced countries in general involve a later vintage technology than older products: this normally means that the associated techniques of production tend to be more capital-intensive, more sophisticated and larger scale.[15] The production of new products in poor countries, and the displacement of old products, will therefore tend to nudge the technology in use towards greater capital intensity and larger scale. This in turn will tend to reduce employment below what it might have been, increase technological dualism, and increase urban concentration.

Thus to complete our classification, we may distinguish between two types of production technology, D, associated with the new products:

D_1 where the production technology is more appropriate[16] (i.e. labour-intensive, small scale, etc.) than that of existing products;
D_2 where the production technology is more inappropriate (capital-intensive, large-scale, etc.) than that associated with existing products.

The total effects of new products on economic welfare will thus consist of the combined effects A → D.

EMPIRICAL EVIDENCE

Difficulties in handling changing product characteristics are notorious and this is one reason why the subject has been neglected. Most statistics classify products into such broad categories that many variations in product characteristics go unnoted. Another reason is the general reluctance of economists to incorporate changes in tastes into welfare analysis. It remains true, as Friedman pointed out in 1972 that, 'economic theory proceeds largely to take wants as fixed'. Lack of empirical attention reflects this theoretical bias. There is an abundance of evidence concerning production efficiency – almost none on consumption efficiency.

Ideally, what is needed are detailed microstudies into the history of product development/introduction in relation to some specified needs in a developing country, describing the changing product characteristics, changing costs per characteristic, the way in which different types of product are marketed and promoted, and the extent of reasons for product displacement over time. Some evidence is provided by existing studies of choice of technique, but mainly incidentally, since for the most part the studies aim to exclude product differences. In this section we summarize the findings of eight studies which throw some light on the question.

Sugar processing

The study by James highlights some of the important differences between the two major forms of sweetener consumed in India.[17] The traditional and unrefined product (gur) is shown to be more nutritious and cheaper than refined sugar although the keeping qualities of the latter are superior. Refined sugar is produced far more capital-intensively in factories. Production on a large scale, in India, began only in 1932–33 with the grant of protection to the industry. As a result, the number of sugar factories increased rapidly from 31 prior to the grant to 111 in 1933–34.[18] Despite the competition from sugar, the share of gur in total consumption of sweeteners has remained stable since then.

The Kenyan soap industry

Langdon's[19] study of the soap industry in Kenya focuses on the changes brought about by the introduction of multinationals (MNCs) in an industry which until the Second World War comprised entirely locally-owned soap factories. The product of the MNCs is more expensive than the hand-processed varieties in addition to being functionally inferior, i.e. it has lower quality ingredients and is less durable in cleaning. These facts notwithstanding, the new products have been highly successful. A major reason for this, in Langdon's view, is the differential promotional expenditures of the MNC subsidiaries and the local firms. The latter devote less than 1 per cent of annual turnover to advertising as opposed to the figure of 6 per cent for the subsidiaries. The shift to the high-income soap products has been at the expense of the growth and profitability of local firms which have been forced on this account to imitate MNC patterns of production. The initial displacement of local products has thus tended to become cumulative. This is turn has 'hurt most consumers in Kenya'.

Bread

Data from Chile on automatic baking revealed an unexpected but probably not unusual pattern. Sliced wrapped bread produced on automatic lines was displacing 'traditional' bread although its price per unit weight was considerably higher. The main advantages of the automatic bakeries were their capacity to advertise and, no doubt, influence consumer tastes, as well as the convenience of the product because of its longer shelf life. The automatic bakeries use very capital-intensive technology embodied in machinery systems which are imported *in toto* from abroad. The smaller bakeries use locally designed and fabricated equipment.[20]

Maize grinding

A study of maize grinding in Kenya[21] found that the product that resulted from modern methods of milling (first introduced into Kenya in 1935) differed considerably from ordinary posho, the output of older methods. The modern refined product was substantially less nutritious; costs of milling were substantially greater (prices charged were 3½ times as great) while methods of production were far more capital-intensive (ten times the investment per worker). None the less, the newer refined product has found a large and growing market. This is in part due to various superior qualities – it keeps better and the taste is preferred by many consumers. In part, it is due to

Table 2.1 Differential between unit nutrient costs for 100 per cent extraction maize flour and locally manufactured breakfast cereals

								Vitamins		
	Unit price differential	*Carbo-hydrates*	*Protein*	*Fat*	*Ash*	*Fibre*	*Calories*	*Thiamine*	*Ribo-flavine*	*Niacin*
Weetabix/maize flour	20	18	19	45	55	22	21	9	3	4
Weetaflakes/maize flour	25	22	30	64	68	29	24	12	3	5
Post Toasties/maize flour	27	22	36	225	164	100	27	18	3	10

heavy promotional expenditure, which is concentrated exclusively on the branded refined product. One enquiry suggested that promotional expenses added as much as one-third to production costs. However, the introduction of the refined product has not displaced the old, which continues to be available, nor has it increased its costs.

Breakfast cereals in Kenya

Kaplinsky[22] finds a 90:1 price differential between the unit cost of the cheapest traditional staple and the most expensive imported cereal: the difference in cost per nutrient is greater because of the superior nutritional content of the traditional food. The differences among locally produced cereals are shown in Table 2.1.

Kaplinsky attributes the consumption of the manufactured breakfast cereals (which are taking an increasing share of the market) to the demonstration effect (consumption which started with European expatriates first occurs among the Kenya elite and then 'moves down') and to promotional expenditure, which is heavy and concentrated among the brand name products. 'Breakfast cereals with high nutrient costs are being aggressively marketed as an alternative to traditional foodstuffs.' The production technology of the advanced-country products is far more capital-intensive than the traditional products. It is estimated that the same investment would create employment of 600 in maize mills compared with 15 in a Weetabix/Weetaflakes plant.

Cement blocks

A study of cement blocks[23] in Kenya found that cement blocks were stronger than the traditional clay blocks. If made with mechanized techniques they were of more standard quality. They cost more (at least twice as much) and used more capital-intensive techniques. There was little promotional expenditure associated with cement block sales – but mechanically produced cement blocks had replaced traditional clay blocks for urban use because of their superior qualities.

Footwear

A study of footwear in Ethiopia[24] found that moulded plastic sandals appear 'to offer much better value as simple covers for the soles of feet for peasant families' [than traditional leather shoes]. The price per pair is less than one-seventh of the hand-sewn leather shoe. The production techniques associated with plastic sandals are far less appropriate than those associated with the production of leather shoes. Fixed capital per employee is over three times as great; the materials are largely imported, while the scale of plant is much larger and employment is concentrated in the formal sector. In contrast leather shoes are often produced on a very small scale by families and use local materials. A study in Kenya,[25] where in 1975 70–80 per cent of shoe consumption was met by the Bata shoe company, found that 'The advantages of capital-intensive mass shoe production have rendered artisan shoe production redundant, because of its high production costs.'

Drugs

The drugs industry provides illustrations of many of the points made in the discussion above.[26]

The development and promotion of new products forms a vital part of the marketing strategy of the large pharmaceutical firms. 'About half of the medicines sold on national markets are less than ten years old.[27] Research and development into new drugs is concentrated in a few large companies in the developed countries. 'Pharmaceutical firms in developing countries do practically no research, with the possible exception of a few leading producers.'[28]

'Foreign firms are not interested in research on drugs for tropical diseases as the global demand for such drugs will not be sufficiently economic.'[29] The consequence of the nature of the industry, its competitive strategy, and the fact that research is concentrated in the developed countries, is that new products tend to be directed at

Table 2.2 Alternative costs of treating tuberculosis

Drug	*Standard adult daily dose*	*Cost per year (US$)*	*Cost as percentage of lowest priced drug*
Isoniazid	300 mg	0.90	–
Thiscetazone	150 mg	1.00	111.1
Paraaminosalicyclicacid	10 g	9.25	1027.8
Streptomycin	1 g	17.25	1916.7

Source: H. Friebel, Therapeutic Needs and Production of Drugs, in UNIDO (1970). *Establishment of Pharmaceutical Industries in Developing Countries: Report and Proceedings of Expert Working Group Meeting* (ID/35), p. 33.

diseases of the rich, rather than the poor, new drugs tend to be high-cost, and product differentiation abounds. While the new drugs usually fulfil some additional benefit, as compared with the old, they often do so at a considerable increase in cost, as is illustrated in Table 2.2. For poorer countries (consumers) the additional benefits may not be justified in relation to the extra costs.

Heavy promotional expenditure in the industry – in the US for example expenditure on advertising exceeds expenditure on research and development by three to four times – raises the cost of each drug, and often misleads as to the relative efficacy of different drugs: 'The flow of essential information about the discovery of drugs, their relative efficiency and their relative cost is submerged in a deluge of glossy and persuasive advertising of particular products.'[30] Lack of official regulation and investigation means that the promotional efforts in LDCs are often potentially more harmful than those permitted in DCs.

Many of the new products do not contain new chemical properties, or therapeutic characteristics, but are differentiated with respect to packaging, presentation, etc. Investigations have established huge cost differences between proprietary branded products and the generic equivalent – e.g. 50–60 times for antibiotics. In the US price differences of up to 1000 per cent have been established.[31] In India it was estimated that 116 drugs represented India's basic drug needs in contrast to the 15 000 drugs currently marketed.[32]

The net result has been drugs which are 'inappropriate in terms of price and therapeutic effects.'[33] 'The high price of modern drugs in developing countries, their heavy promotion and the training received by doctors all conspire with the unequal distribution of income and lack of free health provisions in developing countries to ensure that only a small minority of the population gain access to drugs.'[34] In India, it has been estimated that only 20 per cent of the population consume modern drugs.[35]

SUMMARY OF CASE STUDIES

Table 2.3 summarizes the effects of the introduction of new products in these studies. In only one of the eight – soap – did there appear to be substantial displacement of existing products. For the most part therefore (and this may have been true of the soap case too) the new products extended consumers' choice. Promotional expenditure, however, was probably responsible for over-evaluation of the benefits of the new products and over-consumption in at least six out of the eight cases. This may have had serious negative effects in the food and drug cases. In all the food cases, the nutritional content of the new products was lower than the old. In drugs, the substantial expenditure due to the consumption of expensive new products may have limited expenditure on other drugs. In seven of the eight cases, the new products had high-income characteristics compared with the old. In these seven cases the distributional impact was inegalitarian. Plastic sandals were the one case of a new low-income product whose consumption impact was probably egalitarian. Production technology was inappropriate in every case, including sandals, as compared with the technology associated with the older products.

The net effects on welfare depend on the combined effects on consumption and production. In each of the cases considered there were both positive and negative considerations, so that the overall impact depends on how the various considerations are weighted. The chief positive effect lay in the extension of consumer choice and the benefits conferred on those consumers who prefer the new products to the old; the negative effects arose from over-consumption of the new products due to demonstration/promotion, the inegalitarian consumption effects and the effects on production technology.

It must be emphasized that these products were not selected to be representative of products as a whole. Moreover more detailed and custom-designed research is required in each case to be confident about the results. But the direction of effects is broadly in line with what one would expect, given the historical circumstances. New products could, in principle, be designed for low-income consumers

Table 2.3 Effects of new products

	Benefits some or all consumers		*Negative effects on some or all consumers*		*Distributional consequences*		*Production technology*	
	Reduces costs of characteristics A_1	*Increases choice* A_2	*Displaces old product/ increases cost* B_1	*Promotion/ over-consumption* B_2	*Egalitarian* C_1	*Inegalitarian* C_2	*Appropriate* D_1	*Inappropriate* D_2
Refined sugar	–	+	–	+	–	+	–	+
Soap	–	–	+	+	–	+	–	+
Sliced, wrapped bread	–	+	–	+	–	+	–	+
Refined maize flour	–	+	–	+	–	+	–	+
Breakfast cereals	–	+	–	+	–	+	–	+
Cement blocks	–	+	–	–	–	+	–	+
Plastic sandals	+	+	–	–	+	–	–	+
Drugs	–	+	–	+	–	+	–	+

– indicates negative or zero effect
+ indicates positive effect

to be produced with appropriate technology. But the historical (and current) concentration of research and development into products and techniques in the advanced countries makes this a rarity at present.

SOME POLICY IMPLICATIONS

The introduction into LDCs of new products that have been developed in and for developed countries can have important welfare implications. In many cases, the distributional impact is likely to be inegalitarian, while in some cases, the new products may impose absolute losses on society. Our analysis suggests the need for countries to develop an active product policy. Elements of such a policy include:

1. *Product development.* There is a systematic tendency for product development in high-income countries to be inappropriate for LDCs, leading to products containing excessive characteristics. The development of appropriate products with low-income characteristics (including divisibility) and appropriate to the environment of poor countries would enable such countries to benefit

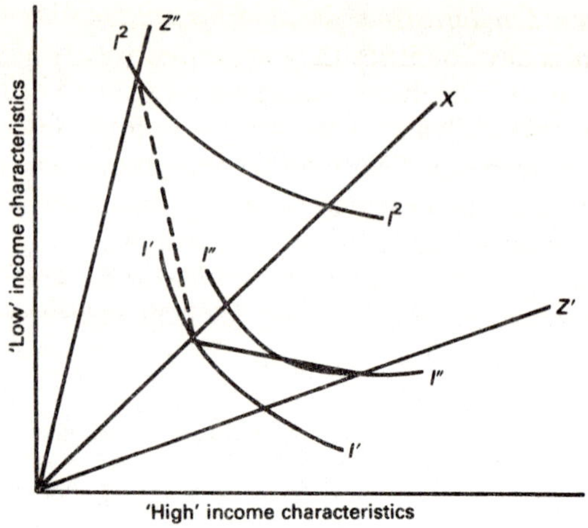

Figure 2.7

from the increased efficiency of modern products, without being harmed by their high-income characteristics. The development of appropriate products should contribute to a more egalitarian income distribution. With an initial inegalitarian distribution, such product development might need to correspond to 'felt needs' of poor consumers, rather than monetary demand. Even where modern products from advanced countries do confer benefits on consumers in poor countries – as in Figure 2.7 – the development of appropriate products would be likely to confer greater benefits.

Suppose X is the initial product. The new product developed in a developed country, Z', enables consumer welfare in the LDC to rise from $I'I'$ to $I''I''$. But more appropriate product, Z'', leads to a greater rise in welfare in the LDC. Policies to develop more appropriate products require local research and development related to product (as well as technique) development.

2. *Policies towards the import of products.* A more selective approach is required towards the import of new products from advanced countries, while a more active search is needed to promote the import of appropriate products from other developing countries. More systematic research is needed to assess the benefits and costs of particular products in particular countries, but it seems likely that there is a strong case for a total ban on some products, and for differential taxation on others.

3. *Government standards/specifications for products need revision*. At present, in many countries they are either totally absent (e.g. no requirements for health warnings on cigarettes) or imitations of requirements in advanced countries, leading to 'too-high' standards and over-specified high-income products. Government standards should be designed to see that basic needs-type characteristics are embodied in the relevant goods.
4. *Policy towards advertising/promotional expenditure*. The disproportionate concentration of promotional efforts devoted to modern advanced country products needs to be corrected. There is a strong case for limitations on promotional expenditure, which wastes resources and creates consumption inefficiency. Paradoxically, while the need for controlled advertising is greater in developing countries than in developed, in most LDCs there are no controls at all. A deliberate effort – of expenditure, organization, publicity – needs to be made to provide accurate information about all available products. Taxation of private promotional expenditure could finance efforts in this direction. The system of trade-marks, which gives rise to much promotional expenditure, often leads to consumer adherence to well-established foreign products. Reform of this system is an important aspect of a products policy, as well as assisting local technological development, and reducing the price of imported technology.[36]
5. *Taxes/subsidies and rationing may be used* to encourage the production and consumption of more appropriate goods, and discourage that of inappropriate goods.

CONCLUSION

This paper has attempted to provide a more systematic framework for assessing the benefits and costs of new products. It presents a research agenda rather than well-established conclusions. We consider the conditions that are likely to lead to gains/losses in consumer welfare following introduction of new products. We find, tentatively, that products developed in advanced countries are likely to have inegalitarian effects when introduced to poor countries and may, under certain conditions, cause losses among some or all consumers. These conclusions suggest the need for LDCs to have an active product policy.

We should emphasize that the paper uses a very conventional model of consumer choice and consumer welfare as its point of departure – viz. it is assumed that consumers are isolated units whose welfare depends on their own behaviour alone and that individual consumer preferences are the ultimate determinant of economic welfare. If we were to introduce interdependencies of behaviour (see Sen) or of consumer welfare (Veblen), or a more sophisticated psychology of consumer welfare (as Scitovsky) we would have to abandon much of the framework presented here.[37] While we have not incorporated any of these approaches in any systematic way, each supports the need to challenge the usual assumption of the unquestioned benefits of new products.

Notes

1. According to Pessemier (1977), in 1970 17 per cent of all sales in manufacturing in the USA consisted of products less than four years old.
2. Midgley (1977), p. 180.
3. Notable exceptions are Veblen and Scitovsky – both have been treated with cold neglect by the profession. See Veblen (1899) and Scitovsky (1976).
4. As Lancaster puts it, 'Introduction of a new good requires that the preference function defined on *n* goods is thrown away and with it all the knowledge of behaviour based on it, and replaced by a brand new function on $n + 1$ goods, or the fiction that the consumer has a potential preference function for all goods present and future – Neither approach gets us very far.' Lancaster (1966), p. 20.
5. Helleiner (1975) has a summary treatment of some of the effects of new products, using Lancaster's approach on similar lines to those adopted in the chapter.
6. An article in the *Harvard Business Review* in 1965 advised the rapid phasing out of products in the final stage of the product cycle on the grounds that they have to be produced in short uneconomical runs and also demand a disproportionate share of scarce managerial and other resources. Kotler (1965).
7. 'Apart from the few weeks at the end of the year when next year's models are already available but dealers also sell out their remaining stock of the old year's models, the consumer must buy his new car with its new design and new features whether he wants them or not.' Scitovsky (1976), p. 256.
8. Research and development on new products is directed at the biggest markets (in monetary terms) as indicated by the figures in Table 2.4 for the pharmaceutical industry.
9. See, for example, Pratten (1971).

Table 2.4 New products introduced by revenue size of therapeutic market, 1969

Therapeutic class revenue parameter $ million	*France*		*Italy*		*Germany*		*UK*	
	No. of C	*No. of P*	*No. of C*	*No. of P*	*No. of C*	*No. of P*	*No. of C*	*No. of P*
0.0–0.5	18	3	17	1	23	2	23	5
0.5–1.5	30	13	34	17	36	19	20	5
1.5–2.5	18	13	19	10	13	14	9	10
2.5–5.0	23	23	29	22	28	47	20	44
5.0–10.0	31	45	27	68	21	41	10	39
10.0+	26	75	22	98	25	112	11	34
Totals	146	172	148	216	146	235	93	137

	Spain		*Belgium*		*Holland*		*Austria*	
	No. of C	*No. of P*	*No. of C*	*No. of P*	*No. of C*	*No. of P*	*No. of C*	*No. of P*
0.0–0.5	55	19	37	3	94	23	116	23
0.5–1.5	43	62	38	21	30	18	28	45
1.5–2.5	19	44	21	8	7	8	4	16
2.5–5.0	27	146	29	21	6	12	–	–
5.0–10.0	7	54	19	24	2	1	–	–
10.0+	6	156	9	21	–	–	–	–
Totals	157	481	153	98	139	62	148	84

C = classes; *P* = products

Source: D. Reekie (1975). *The Economics of the Pharmaceutical Industry*, Macmillan p. 68. Even what relatively little R&D exists in and for poor countries appears to be directed largely to the high-income minorities in these countries. In Berg's survey of Indian firms more than half stated that the new products were aimed at middle- and upper-income levels. Another study notes that in the case of processed foodstuffs very little interest has so far been shown in the lower-income segments of poor countries.

10. A study in the USA covering food and household goods shows for example a distinct trend towards packaging in larger sizes. See Stanton (1964).
11. See Wells (1977).
12. Ibid., pp. 269–70.
13. A similar phenomenon appears to be taking place in Trinidad and Tobago. One-third of households in the poorest four income classes

spend less than 29 per cent of total expenditure on food despite the fact that there is 'considerable undernutrition and malnutrition among the population'.
14. See Hughes and Hunter (1972), p. 90.
15. See discussion in Stewart (1977).
16. There are well-known problems about defining 'appropriate' and 'inappropriate' technology. Despite these it is not difficult to arrive at a working definition (see Stewart, 1977; Morawetz, 1974).
17. James (1977).
18. Government of India (1965).
19. Langdon (1975).
20. Cooper (1976).
21. Stewart (1977), Chapter 9.
22. Kaplinsky (1978).
23. Stewart (1977), Chapter 10.
24. N.S. McBain (1977).
25. Swainson (1978).
26. This example is essentially taken from S. Lall (1975).
27. DAFSA (p. 19).
28. Lall (1975), para. 30.
29. Hathi Committee, p. 93.
30. Lall (1975), para. 87.
31. Kefauver Committee.
32. Hathi Committee.
33. Lall (1975), para. (VI).
34. Ibid., para. 74.
35. Hathi Committee, p. 194.
36. The question of trade marks is explored in much greater depth in UNCTAD (1977) and Stewart (1981).
37. See Sen (1977), Veblen (1899) and Scitovsky (1976).

References

Cooper, C. (1976) *Policy Interventions for Technological Innovation in Less Developed Countries*, Science Policy Research Unit, Sussex University, mimeo.

Documentation et Analyses Financières SA. *The Pharmaceutical Industry in Europe*, 1975.

Government of India (1965). *Report of the Sugar Enquiry Commission*, Delhi.

Hathi Committee on the Indian Drug Industry (1975). *Report of the Committee on Drugs and Pharmaceutical Industry*, New Delhi.

Helleiner, G.K. (1975). 'The Role of Multinational Corporations in the Less Developed Countries' Trade in Technology', *World Development*, Vol. 3, April.

Hughes, C.C. and Hunter, J.M. (1972). 'The Role of Technological Development in Promoting Disease in Africa', in Farvar, M.T. and Milton J.M. (eds), *The Careless Technology*, Natural History Press.

James, J. (1977). *Technology, Products and Income Distribution: A Conceptualization and Application to Sugar Processing in India*, ILO World Employment Programme Research, Nov.

Kaplinsky, R. (1978). *Inappropriate Products and Techniques in UDCS: The Case of Breakfast Foods in Kenya*, Working Paper No. 335, Institute of Development Studies, Nairobi, April.

Kefauver, E. (1966). *In a Few Hands: Monopoly Power in America*, Penguin.

Kotler, P. (1965). 'Phasing Out Weak Products', *Harvard Business Review*, Vol. 43, March–April.

Lall, S. (1975). *Major Issues in Transfer of Technology to Developing Countries: A Case Study of the Pharmaceutical Industry*, UNCTAD, TD/B/C.6/4.

Lancaster, K. (1966). 'Change and Innovation in the Technology of Consumption', *American Economic Review*, Vol. 56, May.

Langdon, S. (1975). 'Multinational Corporations' Taste Transfer and Underdevelopment: A Case Study from Kenya', *Review of African Political Economy*, Jan.

McBain, N.S. (1977). 'Developing Country Product Choice: Footwear in Ethiopia', *World Development*, Vol. 5, Sept.–Oct.

Midgley, D.F. (1977). *Innovation and New Product Marketing*, Croom-Helm.

Morawetz, D. (1974). 'Employment Implications of Industrialisation in Developing Countries: A Survey', *Economic Journal*, Vol. 84, Sept.

Pessemier, E.A. (1977). *Product Management: Strategy and Organisation*, Wiley.

Pratten, C.F. (1971). *Economies of Scale in Manufacturing Industry*, Cambridge University Press.

Scitovsky, T. (1976). *The Joyless Economy: An Enquiry into Human Satisfaction and Consumer Dissatisfaction*, Oxford University Press.

Sen, A.K. (1977). 'Rational Fools: A Critique of the Behavioural Foundations of Economic Theory', *Philosophy and Public Affairs*, Vol. 6.

Speight, A.N.P. (1975). 'Cost Effectiveness and Drug Therapy', *Tropical Doctor*, April.

Stanton W.J. (1964). *Fundamentals of Marketing*, 2nd edn, McGraw-Hill.

Stewart, F. (1977). *Technology and Underdevelopment*, Macmillan.

Stewart, F. (1981). 'International Technology Transfer: Issues and Policy Options', in Streeten, P. and Jolly, R. (eds), *Recent Issues in World Development*, Pergamon.

Swainson, N. (1978). *The Bata Shoe Company: Types of Production and Transfer of Skills*, UNESCO, Division for the Study of Development.

UNCTAD (1977). *The Impact of Trade-Marks on the Development Process of Developing Countries*, TD/B/C/, 6/AC. 33.

Veblen, T. (1899). *The Theory of the Leisure Class*, Macmillan.

Wells, J. (1977). 'The Diffusion of Durables in Brazil and its Implications for Recent Controversies Concerning Brazilian Development', *Cambridge Journal of Economics*, Vol. 1(3).

Appendix
Further Evidence on the Welfare Effects of the Introduction of New Products in Developing Countries

Further evidence on the welfare effects of the introduction of new products in developing countries is derived from two sources. The first (and more detailed) is a series of case studies on product choice and basic needs that were conducted for the International Labour Office and published in 1984.[1] The second is a study by Thomas Biersteker on the impact of multinationals on consumption patterns in Nigeria (which forms part of his larger concern with the role of multinationals in that country).[2] Although not all the evidence from these two studies can be cast in terms of the analytical framework of Chapter 2, each of them tends to confirm the main conclusions reached in that chapter as we shall now attempt to demonstrate.

THE ILO CASE STUDIES

The ILO studies are especially relevant because they are based explicitly on the Lancaster approach to demand and because each case study contains survey evidence of the demand for different products (defined in terms of characteristics) for the industry concerned. Furthermore, in most cases the technological differences associated with the range of competing products were also identified, thus providing a disaggregated picture of the relationships between the demand and supply sides of the question. On the other hand, since the studies were not designed solely to assess the welfare effects of the introduction of new products, they tend to suffer in varying degrees from some of the same limitations as the empirical evidence presented in Chapter 2.[3] The products concerned are soap and detergents, bicycles, furniture, household utensils, footwear and weaning food.[4]

The soap industry in Bangladesh and Ghana

In the soap industries of Bangladesh and Ghana that were studied by Mubin and Forsyth, the same dualistic structures were evident, namely, 'a modern, large-scale, capital-intensive sector, drawing most of its equipment and raw materials from abroad, and a locally oriented small-scale, labour-intensive sector'.[5] As regards consumption patterns as well, they found a similar situation in the two countries: one, that is, where washing soap produced by labour-intensive methods was consumed mainly by the lower-income groups, whereas at higher incomes relatively capital-intensive (washing and toilet) soaps tended to be purchased. However, in Ghana the labour-intensive product was consumed in relatively smaller quantities and Mubin and Forsyth speculate that the same 'rapid elimination of the labour-intensive sectors and the extension of 'modern', urban consumption patterns to the rural areas'[6] may yet occur in Bangladesh as well. 'Certainly, in neither country have the small, labour-intensive producers been able to combat effectively the superior advertising and marketing methods of the large-scale, capital-intensive sector; the combination of irregular access to inputs (and hence irregular delivery of final product) and lack of a brand name were as damaging in Ghana as in Bangladesh.'[7]

Soap and detergents in Barbados

James[8] traces the history of laundry-soap production in Barbados since 1943, the date on which a local firm began production of two brands in bar (and unwrapped) form. He shows how, following the introduction of synthetic detergents in the early 1960s, the production of laundry soap declined (partly because the former possess a major technical advantage over the latter in hard water conditions). In 1973 the local production of laundry soap ceased altogether, though, according to James, there was still a substantial residual demand for the product (especially among the relatively low-income groups). What occurred, he suggests, was a variant of the process whereby an old product is removed by the firm in order to divert demand to its newer version. More specifically, a major shareholder in the original laundry soap firm also owned a large importing agency. 'When a leading international firm manufacturing synthetic detergents appeared as possible clients for the agency, it made obvious business sense for the declining laundry-soap operations to be abandoned.'[9]

In addition to the absolute welfare losses that might thus have been inflicted on them, low-income consumers also seem disadvantaged by the characteristics of the majority of synthetic detergents available in Barbados, which, not only were nearly identical to the products sold in the developed countries (in terms of packaging, marketing and functionality), but appeared to be even more expensive than the synthetic detergents in at least one such country.

Bicycles in Malaysia

Bicycle production in Malaysia dates back to 1959 with the formation of the 'Far East Metal Works'. In 1967 the British firm 'Raleigh' entered the market which remains dominated by these two large-scale manufacturers. Onn's survey results show that the less expensive (and more labour-intensive) 'Far East' brands tend to be purchased more heavily by low-income families, whereas households with medium and high incomes seem to prefer the 'Raleigh' products.[10] Although the former are regarded by Onn as being more appropriate in terms of both product characteristics and production processes, he shows that the latter are favoured by the Malaysian system of industrial incentives and by 'sophisticated' marketing and advertising strategies. Indeed, advertising expenditures per unit of value added by 'Raleigh' are as much as five times as great as those made by 'Far East'.

Furniture in Kenya

House's study of furniture-making in Kenya reveals a similar pattern to the ILO studies that have already been described and to the findings from Chapter 2.[11] In particular:

> The poor households in Nairobi, and by inference in other urban areas and in rural Kenya also, buy low-quality, relatively inexpensive furniture from the very labour-intensive informal sector. As incomes rise, preferences move in favour of more expensive and middle-quality furniture, produced in the small-scale workshops and large-size factories. Both kinds of producers are less labour-intensive than informal-sector producers but they are still very labour-intensive judged against other branches of industry. And both use a very small amount of imported raw materials.
>
> At the top of the income scale, among the prosperous business and professional elite, are found the buyers of the luxury items of furniture. These are the very expensive, highly skill- and import-intensive products.[12]

Such products include 'antique' reproductions or 'ultra-modern' high-quality furniture.[13] Though the amount of advertising undertaken in the industry is reported to be small, what there is of it is undertaken by the sub-sector producing 'high-income' products.[14]

Household utensils in India

Papola and Sinha examine the demand in India for different types of metal utensils that are used, among other purposes, for cooking and storage.[15] Of these, 'Stainless steel is the new and upcoming metal, but it is also the costliest and not all households can afford it.'[16] This metal is also the most capital-intensive of the five alternatives (iron, aluminium, brass, *phool* and stainless steel) that were studied. (On the other hand, aluminium, a modern and relatively capital-intensive metal is also popular among the low-income groups.) Though the demand for stainless steel is growing very rapidly, it does not appear to have displaced the other metals. Indeed, 'No respondent household indicated non-availability in general or in the local market as a factor responsible for their non-fulfilment of requirements or for the absence of metal utensils as an item of current purchase.'[17]

Footwear in Ghana

Aryee's study of footwear in Ghana identifies a wide variety of shoes that differ not only according to type (e.g. sandals versus shoes with heels), but also according to the materials that are used.[18] Predictably, he finds that 'Footwear made wholly or partly of natural leather tends to be more expensive than that made from artificial leather or rubber products.'[19] However, intra-type price differences appeared to be much more pronounced than variations between different types of shoes. This was because 'the price of footwear depends on factors other than natural or artificial leather. For instance, the method of sewing and lasting determines the durability of the footwear. Moreover, in Ghana prices may depend on whether footwear is manufactured and sold in the formal or in the informal sectors.'[20] Surprisingly, Aryee finds no tendency for the poor to buy the relatively inexpensive types of footwear, a finding which he explains with reference to the use pattern that is associated with this household group. In particular, 'Since a higher proportion of footwear owned by the low-income group is used for ceremonial occasions, a higher proportion is also purchased for reasons of fashion and good finishing.'[21] What was more in line with the case studies to which we have already referred, though, was a distinct tendency for the low-income groups to purchase relatively inexpensive versions of a *given* type of footwear from small-scale, labour-intensive firms, and for the higher-income groups to buy instead from large-scale, formal-sector enterprises.

Weaning food in Ethiopia

In Ethiopia the dominant type of weaning food for the vast majority of households remains traditional: that is, food that is similar to what is eaten by adult members of the household.[22] In the urban areas, however, and especially among relatively high-income groups infant formula products are widely used. A low-cost weaning food is produced industrially in Ethiopia (on the basis of machinery imported from and financed by European sources), but even with a subsidy this product appears to have been able to reach only some 5 to 15 per cent of the population. Landgren-Gudina concludes that for the predominantly rural population, a home-made weaning food based on locally available staples and prepared according to recognized nutritional guidelines is probably the most appropriate alternative.

Summary

As was true of the empirical evidence presented in Chapter 2, only in the case of soap (in Barbados) did the introduction of a new product give rise to the displacement of an existing product (with attendant welfare losses for the comparatively large number of consumers who still appeared to favour the latter at the time of its withdrawal from the market). In all the soap cases, as well as bicycles (and to a lesser extent furniture), advertising and marketing may have led to over-valuation or over-consumption of the new products. In general, the new products seemed on the one hand to embody characteristics that were desired mainly by the higher-income groups and on the other hand, to be produced by relatively capital-intensive methods (though some exceptions to this general pattern were observed, including, as in Chapter 2 above, the case of footwear).

THE IMPACT OF MULTINATIONALS IN NIGERIA

In chapter 7 of his book, Biersteker examines the appropriateness of the products that multinationals have introduced in Nigeria (and in particular a sample of products produced in the manufacturing sector). His assessment turns partly on whether these enterprises – as opposed to local firms – were responsible for the introduction of the products into the Nigerian market. In several industries this was apparently not the case: multinationals, that is to say, followed rather than led the process. And where these cases involved inappropriate products, the blame, according to Biersteker, cannot for this reason, be made to rest on the foreign firms. For example:

> All sugar refined in Nigeria is white, fine-granulated refined sugar. This type of sugar is more costly to produce and has a lower nutritional value than the brown, large-crystallized sugar available in some countries. Hence it is possible to argue that the sugar produced in Nigeria is inappropriate. However, the decision to process white, granulated refined sugar was made by a Nigerian governmental firm, not by a multinational corporation intent on altering domestic patterns of production to suit its product.[23]

Thus, although white sugar is deemed to be inappropriate (as it was in Chapter 2) the introduction of it cannot be blamed on the multinational concerned.

Furthermore, the new products that were introduced by the foreign-owned firms were not, according to Biersteker, invariably inappropriate. He cites as an example of this the introduction of synthetic textiles after 1970, which, although being considerably more expensive than cotton textiles made by local firms, are not in his view 'substantively of less value' than the latter. For, although synthetics 'may be much warmer than cotton textiles in a tropical climate like Nigeria's . . . they are also likely to last longer in the tropical sun and with repeated handwashings'.[24] (What this assessment does not provide, however, which would be essential for the product characteristics framework used in Chapter 2, are any measures of the costs per unit of the relevant characteristics, especially the differential durability of the two products. Such estimates may very well show that on balance the new synthetic textiles have an inegalitarian distributional impact in a country such as Nigeria.)

Biersteker does, however, identify numerous cases in which multinationals have introduced inappropriate new products in Nigeria.

> The introduction of powdered baby formulas into the country is one example. Nigerian consumers have been confronted with a barrage of advertising extolling the virtues of using powdered baby formulas in place of breast-feeding. The advertising has emphasized that use of the formulas is more modern than breast-feeding and illustrate this point by portraying upwardly mobile professional woman using the product for their newborn children. . . . A similar argument can be made about food-related products such as breakfast cereals and toothpastes. Both products replace existing products with a more costly commodity of a lower (or questionable) substantive value. Processed breakfast cereals replace porridges made from local grains, and 'modern' toothpastes take the place of the equally (often more) effective chewing sticks popularly used for cleaning and scraping the teeth and gums. [Moreover,] In a few industries multinationals *have* encouraged inappropriate consumption patterns by way of

taste creation . . . Vitamins, skin lighteners, cosmetics, many pharmaceutical and health-related products, and soft drinks marketed by multinationals have no prior substitutes, a considerable economic cost, and a highly questionable substantive value. These industries all employ extensive advertising and marketing techniques to demonstrate that the more modern, stylish, or better-educated population in Nigeria regularly uses these products.[25]

Notes

1. Van Ginneken and Baron (eds) (1984).
2. Biersteker (1978).
3. For a description of the objectives and results of the ILO research see Chapters 1 and 10 of van Ginneken and Baron (eds) (1984).
4. One case study, of passenger transport in Karachi, is not included in our survey of the ILO volume, because it deals only with rather broad product categories such as buses, taxis, bicycles and cars.
5. Mubin and Forsyth (1984), p. 50.
6. Ibid., p. 51.
7. Ibid.
8. James (1984).
9. Ibid., p. 184.
10. Onn (1984).
11. House (1984).
12. Ibid., p. 158.
13. A more detailed discussion of the competing products is to be found in House (1980).
14. Ibid.
15. Papola and Sinha (1984).
16. Ibid., p. 95.
17. Ibid., p. 110.
18. Aryee (1984).
19. Ibid., p. 129.
20. Ibid.
21. Ibid., p. 136.
22. Landgren-Gudina (1984).
23. Biersteker (1978), p. 131.
24. Ibid.
25. Ibid., p. 131–2 (emphasis in original).

References

Aryee, G. (1984). 'Income Distribution, Technology and Employment in the Footwear Industry in Ghana', in van Ginneken and Baron (eds).

Biersteker, T. (1978). *Distortion or Development? Contending Perspectives*

on the Multinational Corporation, MIT Press.
Ginneken, W. van and Baron, C. (eds) (1984). *Appropriate Products, Employment and Technology: Case Studies on Consumer Choice and Basic Needs in Developing Countries*, Macmillan.
House, W.J. (1980). '*Technological Choice, Employment Generation, Income Distribution and Consumer Demand: The Case of Furniture Making in Kenya*', ILO World Employment Programme Research, Working Paper No. 60.
House, W.J. (1984). 'Product Choice and Employment in Furniture-Making in Kenya', in van Ginneken and Baron (eds).
James, J. (1984). 'Technology, Products and Income Distribution: The Soap Market in Barbados', in van Ginneken and Baron (eds).
Landgren-Gudina, M. (1984). 'Weaning Food and Low-Income Consumers in Ethiopia', in van Ginneken and Baron (eds).
Mubin, A. and Forsyth, D. (1984). 'Technology, Employment and Income Distribution: The Soap Industry in Bangladesh', in van Ginneken and Baron (eds).
Onn, F.C. (1984). 'Basic Needs and Appropriate Technology in the Malaysian Bicycle Industry', in van Ginneken and Baron (eds).
Papola, T.S. and Sinha, R.C. (1984). 'Metal Household Utensils and Basic Needs in India', in van Ginneken and Baron (eds).

3 The Adaptation of Developed-Country Products

The main argument of the previous chapter was that new products developed in and for advanced countries tend to be inappropriate to the needs of the majority of those living in developing countries. We argued that, as a result, these products generally have an inegalitarian welfare impact when they are introduced in the Third World.

Implicit in this argument, however, is the assumption that in the transfer process, products from developed countries are not adapted in any way to the conditions of the recipient countries. In what follows we seek to assess the validity of this assumption and, on the basis of the assessment, to determine to what extent the welfare conclusions reached in the previous chapter need to be modified.

THE RELATIONSHIP WITH THE ADAPTATION OF DEVELOPED-COUNTRY TECHNOLOGY

Our concern in this chapter has its counterpart in the debate over the effects of the transfer of technology and, in particular, the possibility that adaption of this technology to the socio-economic circumstances of developing countries might render it more appropriate. In both cases, for example, because the multinational corporation is a major actor in the transfer process, the extent to which it engages in adaptation is necessarily an important component of the issue as a whole (though of course adaptations can be made by other economic factors as well). In both cases, moreover, one can find views on the adaptive behaviour of multinational corporations that are sharply divided, often along ideological lines. Thus, on the techniques side, critics argue that 'because multinational firms do not adapt their production processes, they introduce capital-intensive techniques'.[1] More conventional writers in contrast contend that 'considerable adaptation does indeed takes place'.[2] Similarly, on the products side, critics of the multinational corporation tend to view adaptation as occurring rarely, if ever, while against this it is argued that 'adaptations to specific local conditions and tastes are often made'.[3]

The relationship between the adaptation of products and technologies involves more, however, than merely an analogy, for there may also be an important causal connection between the two. In particular, adaptations of technology may often be dependent on prior alterations in products, in the same way and for the same reason that the choice of technology is known to be crucially dependent on the choice of product.[4] This connection has not gone unnoticed in the technology literature: White, for example, has suggested that 'In many cases adaptation of the processes requires product adaptation, to make fine tolerances less critical and frequently to lower the general quality of the product.'[5] Yet, in spite of the particular significance of product adaptations that this view implies, the technology literature has focused overwhelmingly on adaptations in processes. What few studies deal with the former tend to do so incidentally, rather than as part of a systematic research agenda. Fortunately, the question of product adaptations by multinationals has received more detailed attention in the business and marketing literatures and it is on these that we shall mainly draw in this chapter.

ADAPTATION vs. STANDARDIZATION OF PRODUCT CHARACTERISTICS IN DEVELOPING COUNTRIES

Following the characteristics approach to demand that was adopted in the previous chapter, it is obvious that the decision confronting the multinational of whether or not to adapt products to the developing country environment is a multidimensional one. The extreme possibilities – shown for an illustrative set of characteristics in the first and last columns of Table 3.1 – are those in which characteristics are either all left unchanged (the 'pure standardization' case) or are all adapted (the 'pure differentiation' case).

In the 'pure standardization' case, the multinational operates 'as if the entire world (or major regions of it) were a single entity; *it sells the same things in the same way everywhere*'.[6] In so far as global preferences actually conform (or can be made to conform) to the homogeneous pattern that this type of behaviour implies, global standardization has an obvious appeal, for it avoids all the costs that are associated with the adaptation of product characteristics to particular markets. Coca-Cola and Pepsi-Cola, for example, have been able to capture vast global markets on the basis of exactly the same product characteristics that are used in the United States (in advertis-

Table 3.1 The standardization–differentiation continuum

Characteristics	*Global standardization*	*Mixed strategy*	*Pure differentiation*
Product design	×	×	–
Advertising and marketing	×	–	–
Packaging	×	–	–
Divisibility	×	×	–
Usage instructions	×	–	–
Retail price	×	×	–
Brand name	×	×	–

× = standardization of product characteristic
– = adaptation of product characteristic

Source: Based on S. Douglas and Y. Wind, (1987). 'The Myth of Globalization', *Columbia Journal of World Business*, Winter, Chart 1.

ing costs alone, the Coca-Cola company is said to have saved in this way some $90 million over a twenty-year period[7]).

In a number of other markets as well, such as fast-food (for example, McDonald's), luxury consumer goods (such as Rolex watches and Cartier jewellery), and car-rentals (for example, Avis), multinationals have 'successfully identified global customer segments, and developed global products and brands targeted to these segments'.[8] In yet other cases, even where patterns of demand are relatively heterogeneous across countries, pure standardization may still remain the preferred option for the multinational. This could occur, in particular, where economies of scale in production, R&D and marketing are so pronounced that the resulting cost reductions more than offset the relatively unfavourable conditions on the demand side.

Nearer the other extreme – of 'pure differentiation' – are the operations of Unilever subsidiaries in developing countries. These producers are known to adapt both products and packaging to a variety of developing country circumstances. Food products, for example, are adapted in terms of taste, texture, colour and so on, while packaging is adapted to consumers' needs for size and to climatic conditions in developing countries.[9] On occasion, moreover, Unilever subsidiaries have been known to use as inputs cheap local materials that had previously been disregarded.[10] In part, this heavy reliance on adaptation rather than standardization is dictated by Unilever's product range. Preferences for food in particular tend to be sensitive to cultural differences between developed and developing countries, and attempts to introduce food products from developed to developing countries on a globally standardized basis often encounter precisely this problem.[11] Other Unilever products, such as soaps and detergents, have to contend in addition with various forms of environmental sensitivity, such as water hardness, types of soil and forms of clothing.

On the cost side, as is often the case with producers of packaged consumer goods,[12] production costs in Unilever may be small in relation to marketing costs. This means that there is less incentive to standardize product design than in industries (such as some durable goods) where production costs comprise a relatively high proportion of total costs.

But Unilever's product strategy in developing countries also has a lot to do with its history and the related manner in which the activities of its subsidiaries are organized. In particular, the international activities of the company were oriented from the outset to serving local markets on the basis of manufacturing subsidiaries, which were generally granted a relatively high degree of autonomy.[13] In fact, even Unilever's R&D expenditure is spread over a wide range of developed and developing countries.[14] There is an interesting contrast, in these respects, with the evolution of the Coca-Cola company, which, on the one hand, 'gained true global recognition during World War II, as Coke bottling plants followed the march of US troops around the world'[15] and which, on the other hand, then as now, has a highly centralized decision-making structure.

COMBINING ADAPTATION AND STANDARDIZATION

Although adaptation and standardization are sometimes described as being mutually exclusive, there are in fact a variety of ways in which they can and are combined. One such case (which is illustrated in the second column of Table 3.1) allows some product characteristics to be standardized while others are adapted to the circumstances prevailing in each particular developing country.

For one thing, the way in which the multinational is organized is likely to make it easier for some characteristics to be adapted than others. Standardization of production-related characteristics, for example, will be easier than marketing-related characteristics, when (as is often the case) decisions about the former are made centrally while decisions about the latter are left to the subsidiaries themselves.[16] This same combination of standardized and adapted characteristics will be likely to occur when economies of scale occur in production rather than in marketing, or when cultural differences between countries attach more to the image of the product than to its functional characteristics. Advertising for the Nova car model in Latin America, for example, had to be adapted when General Motors discovered that 'no va' in Spanish means 'does not go'.[17] In another related case, Bournvita and Horlicks are sold in Britain mainly as 'mild soporifics', whereas in Malaysia and India they are sold essentially as energy-giving foods.[18]

Conversely, it may sometimes be the case that product features are adapted while marketing remains essentially unchanged. (A case in point may be the way in which Esso 'adapted its gasoline formulations to meet the weather conditions prevailing in foreign market areas, but employed without change its basic communications appeal, "Put a Tiger in Your Tank".'[19])

Further combinations are made possible by making adaptive decisions for one or more product characteristics along regional lines. 'For example, P & G sells its Pampers brand of diapers worldwide, but the promotional strategy is adapted to different geographic regions.'[20] Similarly, Asian utility vehicles were introduced by some major multinationals in the ASEAN region in the early 1970s,[21] in much the same way that Ford develops different models specifically for its European market. Regional policies for some characteristics, may, of course also be combined with global and national policies for other characteristics. For instance, 'Kellogg's Corn Flakes is sold worldwide, but in some regions such as Latin America, and the Far East, promotional themes are standardized, while in other areas, such as Europe, promotional themes, packaging and distribution strategies are specific to each country.'[22]

Yet another possibility is for the multinational to combine the *columns* of Table 3.1, that is to combine products embodying different bundles of standardized and differentiated characteristics. For example, where economies of scale permit and preferences can be clearly segmented in some way, the multinational may simultaneously market a global brand and a brand that is adapted in certain respects to local circumstances. 'While', for example, 'Philip Morris's and Seagram's country managers and their local advertising agencies are required to implement standard programs for each company's global brands, the managers retain full responsibility for the marketing programs of their locally distributed brands.'[23] Similarly, in a number of subsidiaries in developing countries, Unilever manufactures its international brands of soap for one segment of the population and inexpensive local brands for another segment.[24]

It should by now be apparent that multinationals adopt a wide variety of different policies with regard to the choice between product standardization and adaptation in developing countries. The question that now needs to be addressed is whether any particular pattern of choice can be discerned among these alternatives in the behaviour of large groups of multinationals. The answer to this question in turn therefore requires evidence that has been gathered from *systematic* research efforts and it is to these that we turn next.

SURVEY RESEARCH EVIDENCE

There are not many systematic studies of product adaptations by multinational firms in developing countries. Probably the first of them was undertaken in 1969 by Jack Baranson, as part of a larger study of the automobile industry in developing countries. On the basis of interviews with more than 50 manufacturers of automobiles and parts, Baranson concluded that, 'In transferring automotive production to developing countries, international firms have kept adjustments in product design and production techniques to a minimum. This is because such adjustments are costly and disrupt the industrial transplant process.'[25] This finding suggests that examples of automobile adaptations in developing countries, such as the ASEAN utility vehicle cited earlier, may be very much exceptions to the general rule.[26] On the other hand, Baranson's conclusion would seem to be less valid nowadays, in so far as technical change in microelectronics has reduced the importance of economies of scale in automobile production and has thereby made it easier to accommodate differences in preferences within and between countries.[27]

Reuber's influential 1973 study of the impact of multinationals on developing countries incorporates a much broader range of industries.[28] His survey of 77 firms in those industries showed, first, that 20 of them had made some adaptations to the product design. As one would expect, such adaptations were least evident in export-oriented types of foreign investment and most apparent in projects that were initiated by the government of the host country. Reuber's survey also revealed that adaptations in marketing methods were much more prevalent than adaptations in the product design. In particular, 'In only eighteen out of the fifty-eight projects on which information was provided did the investing firms introduce the marketing methods followed at home with no adjustment.'[29]

The usefulness of these findings, however, is limited by their aggregative character – only two forms of adaptation are considered – and by the fact that no information is provided about the nature of the adaptations that took place. Fortunately, a much more detailed study is available that does pay some attention to these limitations of Reuber's results. This study was conducted by John Hill and Richard Still and published in the *Harvard Business Review* in 1984.

These authors surveyed 61 subsidiaries of (mostly) American multinationals producing packaged consumer goods in 22 developing countries. Representatives of these companies were asked about the adaptation of products that had first been marketed in the United States or in Britain and which had subsequently been transferred to developing countries. More specifically, Hill and Still sought information about which of nine product characteristics (measurement units, package size, labelling, package aesthetics, product constituents, package protection, product features, usage instructions and brand name), had been adapted and about the reasons underlying any such changes. Their results, which, in total, cover 174 products, are shown in Table 3.2.

The columns of the table show the nine characteristics in question, while the first row provides information about the overall extent to which each of these characteristics was in fact adapted by the sample firms. It shows, for example, that in 53 per cent of the cases, package sizes were adapted to developing country conditions. The subsequent rows of the table are made up of the explanatory variables and they are divided into two groups, namely, 'environmental' and 'marketing' factors.

Each cell of the table contains two entries: percentages at the bottom are read vertically and measure the explanatory contribution made by a particular variable to the adaptation of a given product characteristic. For example, adaptations of measurement units are caused primarily by legal considerations (44.1 per cent) and consumer preferences (41.5 per cent). The entries at the top of each cell are read horizontally and measure the differential degree of influence of a given explanatory variable across the nine product characteristics. Whereas, for example, purchasing power differences between developed and developing countries bear exclusively (100 per cent)

Table 3.2 The nature and extent of product adaptations

Factors causing product changes	*Measurement units*	*Package sizes*	*Labelling*	*Package aesthetics*	*Product constituents*	*Package protection*	*Product features*	*Usage instructions*	*Brand name*	*Number of times factors were cited as causing adaptations with percentage of total*
	70%	53%	69%	54%	58%	35%	33%	23%	31%	
Environmental factors										
Legal considerations	45.8% 44.1%		34.6% 34.0%		7.2% 5.6%	1.3% 2.3%			11.1% 30.9%	100.0% 153 13.8%
Literacy and education			25.8% 5.1%					38.7% 30.7%	35.5% 20.0%	100.0% 31 2.8%
Purchasing power differences		100.0% 19.9%								100.0% 31 2.8%
Socio-cultural customs and taboos	8.3% 1.2%		33.3% 5.1%	20.8% 3.7%	12.5% 1.5%		8.3% 1.6%		16.8% 7.2%	100.0% 24 2.1%
Climate		7.2% 0.6%				92.8% 14.7%				100.0% 14 1.3%
Total number of environmental factors Percentage of total factors	72 45.3%	32 20.5%	69 44.2%	5 3.7%	14 7.1%	15 17.0%	2 1.6%	12 30.7%	32 58.1%	253 22.9%

Marketing factors										
Consumer preferences	18.8% 41.5%	9.1% 20.5%	13.9% 31.4%	20.2% 53.0%	15.6% 27.9%		12.2% 34.9%	4.5% 41.1%	5.7% 36.3%	100.0% 352 31.8%
Competition		18.9% 19.9%	19.5% 20.5%	21.9% 26.9%	17.8% 14.8%		21.3% 28.5%		0.6% 1.8%	100.0% 164 14.8%
Material availability					54.3% 41.7%	21.2% 36.3%	24.5% 30.2%			100.0% 151 13.6%
Consumer purchasing habits		76.3% 37.2%			9.2% 3.5%	14.5% 12.5%				100.0% 76 6.9%
Changes in distribution						100.0% 29.5%				100.0% 26 2.3%
Total number of marketing factors Percentage of total factors	66 41.5%	124 79.5%	81 51.9%	107 79.9%	173 87.9%	69 78.4%	115 93.6%	16 41.1%	21 38.2%	769 69.4%
Other factors Percentage of total factors	21 13.2%	3 1.9%	6 3.9%	22 16.4%	10 5.0%	4 4.5%	6 4.8%	11 28.2%	21 3.6%	85 7.7%
Total number of factors	159.0 100.0%	156.0 100.0%	156.0 100.0%	134.0 100.0%	197.0 100.0%	88.0 100.0%	123.0 100.0%	39.0 100.0%	55.0 100.0%	1 107.0 100.0%

Source: Hill and Still (1984), Exhibit I.

on adaptations in package sizes, consumer preferences influenced the adaptations of all but one product characteristic (namely, package protection).

It is clear that these findings contradict the view held, as noted earlier, by critics of the multinational, that products from developed countries are seldom, if ever, adapted to Third World conditions. For, in five of the nine columns of Table 3.2, multinationals adapted their products on more than half the total number of occasions that were studied by Hill and Still, and in only one case (that of usage instructions) was the rate of adaptation less than 30 per cent; however, in interpreting these results it is important to bear in mind that the products covered by the study include food and drink, as well as other products used inside the home which are particularly sensitive to cultural differences between countries and which are therefore particularly prone to adaptation.[30] It is also clear from the examples provided in the text of the article that some of these adaptations, such as translations of labels and measurement units into local terminology benefit consumers in general, compared to the products that are transferred unmodified into developing countries.

What is much less clear, though, is whether and to what extent the findings reported by Hill and Still challenge the main conclusions that were reached in the previous chapter. For what was argued there was only partly about the *absolute* gains that are derived from new products. In addition, and more importantly, the argument concerned the *relative* benefits from these products and in particular what we described as the skewed distribution of benefits between high- and low-income consumers. From this point of view, the impact of adaptations depends on:

1. the extent of adaptation across different characteristics and whether in particular there is any tendency for the characteristics that are most extensively adapted to be those that are regarded as being especially important to any particular income group;
2. whether for any given degree of adaptations, the directions of change favour the preferences of one income group rather than another.

Consider first condition (1). For most of the characteristics that comprise the columns of Table 3.2, a classification into high- and low-income categories is extremely difficult, which is not surprising, given the fact that the survey was not designed for this purpose.

However, two characteristics, package aesthetics and product features, stand out as exceptions. The former can reasonably be regarded as a 'high-income' characteristic, while the latter crucially determines the extent to which adaptations alter the balance of characteristics that is normally associated with developed-country products (a balance that, following our argument in the previous chapter, generally tends to favour 'high-', over 'low-income' characteristics). A low rate of adaptation of this latter characteristic accordingly suggests a lack of countervailing influence in favour of 'low-income' characteristics (though, of course, a high rate of adaptation does not in itself establish the existence of any such influence). Table 3.2 shows that this is indeed the case: product features are adapted in only one case out of three, whereas package aesthetics in contrast are altered in nearly twice as many instances, a result, we may note, that is essentially very similar to what Reuber found. For these characteristics at least, therefore, the distributional impact does not appear to depart from the inegalitarian pattern that was identified in the previous chapter.

Consider next the direction of adaptations, which, for a given rate of change, determines the extent of benefits that are conferred on different groups of consumers. This is in fact a distributional question of exactly the same kind that we addressed in the previous chapter in relation to new products. For a number of reasons, however, the Hill and Still results do not permit a very clear answer to this question. For one thing, the explanatory variables that comprise the rows of the table reflect only the motivations underlying adaptive behaviour and these do not of course always correspond to actual outcomes. An intention to adapt products according to the preferences of poor consumers, for example, may turn out in practice to have a very different impact. Another difficulty is that the explanatory variables cannot all be divided into categories that reflect the interests of different income groups. The distributional impact of product adaptations that are ascribed to 'competition' or 'material availability', for example, is entirely indeterminate.

It is nevertheless very striking that most forms of adaptation reveal very little sensitivity to the variables – purchasing power differences, literacy and education and socio-cultural customs – that would seem, at least partly, to determine the extent to which benefits are conferred on low-income consumers. These three variables are in fact respectively able to explain only 2.8, 2.8 and 2.1 per cent of the total number of adaptations reported in Table 3.2. One exception to this general pattern is packaging size, 20 per cent of the adaptations of which are explained by differences in purchasing power. Since the rate of adaptation of this characteristic is itself high (at 53 per cent), it clearly warrants our attention. Package size can in fact be regarded less as a 'low-income' characteristic, than as a characteristic that determines the degree of divisibility of new products. The previous chapter suggested that products designed for high-income consumers tend to be more indivisible than would be suitable for low-income consumers and that, as a result, new products benefit the rich rather than the poor. Hill and Still's findings, however, appear to challenge this conclusion. In particular, 'In LDCs such products as cigarettes, razor blades and chewing gum are sold in small or individual units so that consumers with limited discretionary incomes can afford to buy them. Warner Lambert, which sells Chiclets chewing gum in 12-piece packs in the United States, decided to sell 2-piece packs for the equivalent of a few cents each in Latin America.'[31] Closely related to package size is the adaptation of package protection, which, as Hill and Still point out, on occasion also benefits the poor. 'Less affluent consumers in Mexico, for instance, cannot afford to buy detergents in large packages, and manufacturers find cardboard too expensive for small detergent packages. Smaller units are therefore put into plastic bags.'[32]

A second exception to the generally low degree of sensitivity of adaptations to the three variables mentioned above is the case of usage instructions. Adaptations of this characteristic, which, as Table 3.2 shows, are explained in relatively large part by literacy and education, are closely related to what in the previous chapter was described as 'learning by consuming'. It was suggested there that a systematic tendency exists for inefficiency of consumption of new products in poor countries. The finding reported in Table 3.2 that usage instructions are adapted by multinationals in only one out of five cases tends to confirm rather than contradict this point of view. For, without any alteration, usage instructions that are appropriate to developed-country consumers are often likely to be quite unsuitable for uneducated rural consumers in developing countries. However, as Hill and Still emphasize, what adaptations *are* made to this product characteristic often benefit precisely these groups. They point, for example, to the way that 'Instructions are screened for terms unfamiliar to less advanced consumers.'[33] In addition, 'Consumer goods companies in South-East Asia and Southern Africa use movies to advertise and demonstrate products. Managers arrange to show popular films in suburban or rural areas. In between movies, commercials – often 10 to 20 minutes long – show what detergent or toothpaste is, how to use it, and the benefits consumers can expect from its use.'[34]

In the case of both package sizes and usage instructions, however, the extent to which low-income groups might have benefited from the product adaptations that occurred depends importantly on exactly how one interprets the variable 'consumer preferences'. For in both cases, and especially in the case of usage instructions, Table 3.2 shows that this variable accounts for a relatively high percentage of all adaptations. Indeed, for all product characteristics taken together, consumer preferences account for almost one third of all adaptations. Unfortunately, because Hill and Still provide little indication of exactly *whose* preferences are included in this variable, no distributional implications can be directly determined.

It is difficult, though, to conceive of much participation by the poor in the benefits of adaptation, to the extent that multinationals emphasize product differentiation rather than price competition. For an emphasis on the former, by allowing the multinational to extract a price premium, effectively limits the degree to which adaptations in product design can actually become accessible to the poor. Precisely this problem has made it difficult, for example, to reach large numbers of the poor 'with proprietary foods marketed in the conventional manner'.[35] And the major reason for this is seen as 'the inability to reconcile the demand for corporate profit with a product low enough in cost to reach the needy in large numbers'.[36]

How prevalent then is product differentiation based on brand promotion by multinationals operating in the Third World? A number of economists argue persuasively that because this form of (imperfect) competition is the very factor on which the competitive advantage of the multinational rests, it becomes almost a defining feature of the operations of these firms. Wells, for example, has suggested that for such products as simple consumer goods 'the competitive advantage that US and similar firms have exploited has typically been in brand development. Once that advantage is abandoned, the firms may have no special skills and enable them to survive abroad.'[37] An essential part of retaining this advantage is of course advertising and marketing, and research conducted by Connor at three different levels of aggregation indicates that the intensity of these efforts in the developing countries he studied is very similar to what is found in the United States.[38] His conclusion is that 'This "exportation" of product differentiation patterns probably results because MNCs tend to implement modes of rivalry they are most familiar with and have found most successful in the industries of their "home" country, the United States in this case.'[39]

There is little in the evidence cited by Hill and Still to suggest a movement away from this description of how the multinational operates. Indeed, the strong resistance to changes in brand names that they report only seems to confirm the importance of this factor. Equally, the adaptations in packaging and advertising that they describe (such as the choice of 'colors that suggest joy, pleasure, and personal well-being in local cultures'[40]), would seem to reflect changes in the manner in which products are differentiated rather than an alteration in this form of competition itself.

Yet, for all the influence that brand promotion apparently exerts, it is not entirely unknown for the multinational to dispense with the strategy of product differentiation, for at least some of its products in Third World markets. Consider, for example, Unilever, which was one of the companies included in the survey by Hill and Still and which we referred to earlier as an example of a multinational that engages relatively heavily in product adaptations. On the one hand, as the very detailed study by Fieldhouse clearly shows, this company preferred in general to market brand differentiated products similar to those designed for developed countries. For, with 'Other things being equal, as Unilever men constantly remarked, the most rewarding market was one that could absorb the highest proportion of branded products which sold on consumer preference and could therefore command a premium, as contrasted with one where it was necessary to sell unbranded goods that sold mainly on price.'[41] Yet, clear as it was, this preference did not always mean that Unilever had cause to eschew the more 'anonymous' and less profitable types of products. For while internationally known brands dominated production from the outset in some Unilever subsidiaries (for example, in Sri Lanka), in other plants in Africa, production began in the 1930s and 1940s (sometimes in local firms that had been purchased by the company) on the basis of simple and inexpensive bar soaps, which, at the time, seemed all the limited local markets could bear. (In Nigeria, for instance, a red carbolic soap was made from 85 per cent palm oil and 15 per cent resin.[42]) As incomes rose in these markets, Unilever began to introduce its well-known international brands (such as Lux and Lifebuoy), but production of the original type of soap often continued nevertheless. Apparently, managers felt that it was more profitable to continue selling the relatively unprofitable products than to discontinue production, in the hope that demand could be shifted to the branded products. In one African case, for example, it was suggested that the 'policy of selling cheap soaps at very low profit margins would check competition and the real profits could be made on quality soaps sold at a premium.'[43]

A COMPARISON WITH PRODUCT ADAPTATIONS BY MULTINATIONALS IN DEVELOPED COUNTRIES

Does the adaptive behaviour of multinationals that has been described in this chapter differ significantly from the behaviour of (comparable) multinationals that operate in the developed countries? While there is no known study of these countries that can be directly compared with the evidence cited above, a survey of twenty-seven American and European multinationals operating in consumer packaged goods industries in Europe and the United States is broadly comparable with the Hill and Still survey.[44]

These studies are comparable because both deal with adaptations of a range of product characteristics; both cover many of the same industries and even many of the same companies. However, the results of the two studies are not directly comparable, partly because they are based on a somewhat different methodology. In particular, whereas Hill and Still present their results in terms of just two categories (namely whether product characteristics are standardized or adapted), the other study employs three categories (namely, whether characteristics are standardized to a high, moderate or low degree). Moreover, the product characteristics with which the two studies are concerned are not in all cases themselves directly comparable.[45]

Nevertheless, there is a rather striking similarity in the main conclusions that are reached: in each case, the product design and brand name are among the three most standardized of all the characteristics studied. For the developing countries this was apparent from Table 3.2 above, which revealed that in the majority of cases product features and brand names are not adapted. For the developed countries, Sorenson and Wiechmann found that 'product characteristics' fell into the 'highly standardized' category in 81 per cent of the total number of cases, while with brand names, no less than 93 per cent of all cases fell into this category.

What this seems to suggest is that the same underlying factors may be at work in influencing the differential extent of adaptation of product characteristics by multinationals in both developed and developing countries. That is, the same factors that make it difficult for multinationals to adapt product designs and brand names in developing countries, may well also make it difficult for them to do so in developed countries. And these factors appear to exert a powerful influence even in industries that (as noted above) are especially sensitive to cultural differences between countries.

CONCLUSIONS

This chapter has been concerned mainly with two questions, both of which relate to the welfare effects of new products that were described in the previous chapter. The first question is whether and to what extent multinationals adapt the characteristics of the products that they transfer to developing countries and the second concerns the distribution of the benefits from any such adaptations.

In relation to the former, the main conclusion to be drawn from the survey evidence is that while multinationals frequently adapt some product characteristics to local conditions, the basic product design remains unchanged in the majority of cases (even in relation to products that are considered especially prone to adaptation). This finding is important because without fundamental design adaptations, new products are unlikely to embody characteristics in proportions that are relevant to the majority of those living in the Third World. Much less clear conclusions, however, can be drawn in relation to the distribution of the benefits from the adaptations of product characteristics that did occur, chiefly because the survey evidence does not address itself to this question. Our impression of what evidence there is, is that while some adaptations have been beneficial to the poor, on the whole this group has not gained much from the adaptive process.

Incomplete though it is, we therefore see little in the available evidence of product adaptations to doubt what in the previous chapter was described as an inegalitarian distribution of the benefits of new products in developing countries.

Notes

1. Biersteker (1978), p. 56.
2. Ibid.
3. Jenkins (1988), p. 1363.
4. On the connections between choice of product and choice of techniques, see Stewart (1977).
5. White (1978), p. 51.
6. Levitt (1983), p. 92–3 (emphasis added). Levitt argues that multinationals *should* behave in this way, to accommodate what he sees as the convergence of global preferences (caused by technical improvements in communications, transport and travel), to benefit from the economies of scale that these convergent preferences permit. Levitt's thesis, however, has been heavily criticized in the business and marketing literature. See, for example, Douglas and Wind (1987) and Onkvisit and Shaw (1987).
7. Austin (1990).
8. Douglas and Wind (1987), p. 21.
9. Veldhuis (1979).
10. Ibid.
11. In one example cited by Austin (1990), a major American food producer attempted to introduce its top-selling American brand into the Thai market without any modifications. The result 'was an expensive failure, because consumers rejected the product's taste; the product formulation was not culturally transferable' (ibid., p. 299).
12. Takeuchi and Porter (1986).
13. Fieldhouse (1978).
14. Veldhuis (1979).
15. Quelch and Hoff (1986), p. 59.
16. Ibid.
17. Kotler (1984).
18. Medawar (1979).
19. Keegan (1969), p. 60.
20. Douglas and Wind (1987), p. 28.
21. Watanabe (1981).
22. Douglas and Wind (1987), p. 28.
23. Quelch and Hoff (1986), p. 68.
24. This is discussed further below.
25. Baranson (1969), p. 24.
26. A more recent review of the automobile industry in developing countries comes to the same conclusion. Thus, 'only token efforts have been made by the transnational firms to develop vehicles that are designed for the particular needs of developing countries.' See Bennett and Sharpe (1985), p. 224.
27. See, for example, Mody and Wheeler (1990).
28. Reuber (1973).
29. Ibid., p. 199.
30. See Austin (1990), Quelch and Hoff (1986) and Takeuchi and Porter (1986).
31. Hill and Still (1984), p. 95.

32. Ibid., p. 98.
33. Ibid., p. 100.
34. Ibid., p. 100.
35. Berg (1973), p. 158.
36. Ibid., p. 158.
37. Wells (1983), p. 63.
38. Connor (1977).
39. Ibid., p. 109. Connor also found 'modest' R&D expenditures by multinationals in Brazil and Mexico; these were much less than in the United States.
40. Hill and Still (1984), p. 95.
41. Fieldhouse (1978), p. 573.
42. Ibid.
43. Ibid., p. 113.
44. Sorenson and Wiechmann (1975).
45. Hill and Still, for example, consider both packaging 'aesthetics' and packaging 'protection' whereas Sorenson and Wiechmann deal only with 'packaging'.

References

Austin, J. (1990). *Managing in Developing Countries: Strategic Analysis and Operating Techniques*, Free Press.

Baranson, J. (1969). *Automotive Industries in Developing Countries*, World Bank Staff Occasional Papers, No. 8.

Bennett, D.C. and Sharpe, K.E. (1985). 'The World Automobile Industry and its Implications', in Newfarmer, R.S. (ed.), *Profits, Progress and Poverty*, University of Notre Dame Press.

Berg, A. (1973). *The Nutrition Factor*. Brookings Institution.

Biersteker, T. (1978). *Distortion or Development? Contending Perspectives on the Multinational Corporation*, MIT Press.

Connor, J. (1977). *The Market Power of Multinationals: A Quantitative Analysis of U.S. Corporations in Brazil and Mexico*, Praeger.

Douglas, S. and Wind, Y. (1987). 'The Myth of Globalization', *Columbia Journal of World Business*, Winter.

Fieldhouse, D. (1978). *Unilever Overseas: The Anatomy of a Multinational Enterprise 1895–1965*, Croom Helm.

Hill, J. and Still, R. (1984). 'Adapting Products to LDC Tastes', *Harvard Business Review*, March–April.

Jenkins, R. (1988). 'Transnational Corporations and Third World Consumption', *World Development*, Vol. 16 (11).

Keegan, W. (1969). 'Multinational Product Planning: Strategic Alternatives', *Journal of Marketing*, Jan.

Kotler, P. (1984). *Marketing Management*, Prentice-Hall.

Levitt, T. (1983). 'The Globalization of Markets', *Harvard Business Review*, May–June.

Medawar, C. (1979). *Insult or Injury? An Enquiry into the Marketing and*

Advertising of British Foods and Drug Products in the Third World, London, Social Audit.

Mody, A. and Wheeler, D. (1990). *Automation and World Competition: New Technologies, Industrial Location, and Trade*, Macmillan.

Onkvisit, S. and Shaw, J. (1987). 'Standardized International Advertising: A Review and Critical Evaluation of the Theoretical and Empirical Evidence', *Columbia Journal of World Business*, Fall.

Quelch, J. and Hoff, E. (1986). 'Customizing Global Marketing', *Harvard Business Review*, May–June.

Reuber, G. (1973). *Private Foreign Investment in Development*, Clarendon Press.

Sorenson, R. and Wiechmann, U. (1975). 'To What Extent Should a Consumer Goods Multinational Corporation Vary its Marketing From Country to Country?', *Harvard Business Review*, May–June.

Takeuchi, H. and Porter, M. (1986). 'Three Roles of International Marketing in Global Strategy', in Porter, M.E. (ed.), *Competition in Global Industries*, Harvard Business School Press.

Veldhuis, K. (1979). 'Transfer and Adaptation of Technology: Unilever as a Case Study', in Robinson, A. (ed.), *Appropriate Technologies for Third World Development*, Macmillan.

Watanabe, S. (1981). 'Multinational Enterprises, Employment and Technology Adaptations', *International Labour Review*, Vol. 120 (6), Nov.–Dec..

Wells, L. (1983). *Third World Multinationals*, MIT Press.

White, L. (1978). 'The Evidence on Appropriate Factor Proportions for Manufacturing in Less Developed Countries: A Survey', *Economic Development and Cultural Change*, Vol. 27.

4 Galbraith Revisited: Advertising in Non-Affluent Societies*

With Stephen Lister

I. INTRODUCTION

Advertising has been debated by economists almost exclusively in the context of developed countries (DCs). We argue here the case for studying advertising in the specifically less-developed country (LDC) context. Our view is that *both* sides in the DC debate have tended to presume that there is no 'problem' of advertising in LDCs. On the one hand, for those who hold that advertising is not a problem in DCs it is natural to assume by extension no such problem in LDCs either.[1] On the other hand, J.K. Galbraith, who is the leading proponent of the view that advertising is crucial to an explanation of the dynamics of advanced capitalist societies, has also implied that the problem is exclusively one for DCs. We show in part II of the paper that this aspect of the Galbraithian view is not supported by the facts. We go on in part III to consider the ways in which the practical effects and welfare implications of advertising may differ in degree or kind in the LDCs. It should be stressed that our aim is not to assess whether advertising is on balance desirable or not for LDCs – in any case an absurdly broad question – but to show with what data are available, that there is a strong case for further research on its specific effects.

II. THE GALBRAITHIAN VIEW

Galbraith has consistently derided the traditional notion of consumers' sovereignty – the view that production is subordinate to exogenously determined demands. Producers, he argues, 'bring into being wants that previously did not exist'.[2] Galbraith however, is not particularly troubled by the welfare implications of abandoning con-

* We are grateful to G.K. Helleiner and C. Medawar for comments on an earlier draft.

sumer sovereignty since he believes that it is only in affluent societies that sovereignty is in fact transferred to producers. Thus:

> We should always be careful not to over-look the obvious. The fact that wants can be synthesized by advertising, catalyzed by salesmanship, and shaped by the discreet manipulations of the persuaders shows that they are not very urgent. A man who is hungry need never be told of his need for food. If he is inspired by his appetite, he is immune to the influence of Messrs. Batten, Barton, Durstine and Osborn. The latter are effective only with those who are so far removed from physical want that they do not already know what they want. *In this state alone men are open to persuasion.*[3]

Or again:

> The opportunity for product differentiation – for associating monopoly power with the brand or personality of a particular seller – *is almost uniquely the result of opulence*[4] [and] the need and the opportunity to persuade people arise only as people have the income to satisfy relatively unimportant wants, of the urgency of which they are not automatically aware.[5] [Indeed, in lower-income societies,] all the commercial advantages lie with the producers of plain bread, sidemeat and oatmeal[6] [where, furthermore,] the tendency for other forms of commercial rivalries, as substitutes for price competition, to be channelled into advertising and salesmanship would disappear.[7]

In effect, therefore, Galbraith asserts: (1) that advertising is minimal in poor countries, and (2) that levels of advertising expenditure ought to be strongly correlated with levels of affluence.[8]

The refutation

We tested both these propositions using 1974 data for the 57 DCs and LDCs shown in Table 4.1. The countries are ranked according to *per capita* national incomes and the final column shows advertising as a percentage of national product.

It is immediately apparent that the proportion of national product represented by advertising is significant in many poor countries. Two of the poorest countries listed – Kenya and Thailand – devoted about half of 1 per cent of national product to advertising (i.e. roughly half the average of the DCs). Moreover, many developing countries of

Table 4.1 Expenditure on advertising, 1974

Country	Total ($m)	Print (%)	Cinema Radio TV (%)	Other (%)	National product (US dollars)	As % of GNP
1. Kuwait	9.7	33	44	23	10 863	0.10
2. Switzerland	744.2	41	5	54	6 790	1.70
3. Sweden	506.0	50	2	48	6 153	1.01
4. Australia	864.3	49	39	12	5 986	1.08
5. Canada	1 703.7	41	24	35	5 962	1.27
6. USA	26 750.0	40	25	35	5 922	2.13
7. Austria	342.6	35	19	46	5 892	1.17
8. W. Germany	2 523.1	52	16	32	5 470	0.74
9. Denmark	355.3	60	1	39	5 357	1.31
10. Belgium	276.9	55	3	42	5 028	0.56
11. Norway	178.2	78	2	20	4 953	0.90
12. Netherlands	771.8	60	8	32	4 786	1.19
13. France	1 981.7	35	13	52	4 510	0.84
14. Finland	256.9	53	11	36	4 241	1.29
15. New Zealand	119.4	71	25	4	4 058	0.97
16. Israel	65.0	78	9	13	3 570	0.55
17. Japan	4 163.4	39	38	23	3 569	1.06
18. UK	2 215.0	68	23	9	3 116	1.27
19. Italy	942.0	50	25	25	2 479	0.69
20. Spain	753.1	47	24	27	2 276	0.94
21. Ireland	39.7	59	36	5	2 047	0.63
22. Venezuela	124.3	39	55	6	2 046	0.52
23. Greece	45.4	40	56	4	2 019	0.25
24. Argentina	285.0	44	35	21	1 876	0.60
25. Portugal	54.7	36	49	15	1 442	0.43
26. Iran	47.6	43	43	14	1 371	0.84
27. Cyprus	2.3	30	61	9	1 293	0.28
28. South Africa	356.2	59	15	26	1 147	1.25
29. Malta	1.6	44	56		1 132	0.44
30. Jamaica	17.6	49	42	9	1 090	0.81
31. Mexico	353.4	26	70	4	1 020	0.60
32. Panama	8.9	20	70	10	992	0.55
33. Brazil	941.9	23	34	43	958	0.94
34. Surinam	1.1	36	64		910	0.29
35. Costa Rica	12.8	36	61	3	798	0.83
36. Malaysia	36.6	84	16		715	0.44
37. Mauritius	1.4	43	43	14	632	0.25
38. Dominican Republic	9.1	15	85		583	0.34
39. Syria	8.2	18	15	67	547	0.21
40. Zambia	9.8	42	19	39	524	0.39
41. Rhodesia	12.9	54	44	2	505	0.42
42. Ecuador	23.4	31	41	28	478	0.70
43. Colombia	81.7	20	53	27	477	0.71
44. Peru	59.5	14	60	26	468	0.83
45. South Korea	85.9	42	50	8	436	0.59
46. Ghana	2.4	58	21	21	394	0.06
47. Philippines	50.3	39	52	9	324	0.37

Table 4.1 continued

Country	Total ($m)	Print (%)	Cinema Radio TV (%)	Other (%)	National product (US dollars)	As % of GNP
48. Liberia	0.6	17	67	16	321	0.11
49. Thailand	71.2	11	42	47	304	0.57
50. Egypt	39.4	77	8	15	263	0.20
51. Sri Lanka	2.2	59	36	5	228	0.07
52. Kenya	10.7	40	28	32	197	0.42
53. Indonesia	40.0	46	16	38	175	0.18
54. Pakistan	13.4	16	43	41	154	0.13
55. India	96.4	56	16	28	136	0.12
56. Nepal	0.6	50	17	33	98	0.05
57. Ethiopia	0.7	29	71		91	0.03

Sources: *Advertising Age* (29 November 1976), p. 74; *UN Statistical Yearbook* (1977), Table 193.

Central and South America (viz. Brazil, Costa Rica, Peru, Jamaica, Ecuador and Colombia) spent at least as high a proportion of national product on advertising as several highly developed countries (Belgium, West Germany, Italy, Ireland and France).

Consequently, it is no surprise to find that regressing *per capita* national income on advertising as a percentage of national product yields a rather low positive correlation, with an $\bar{R}^2$ of only 0.31.[9]

Another way of viewing comparative levels of advertising expenditures is to express these as a proportion of manufacturing product. These estimates for a smaller group of LDCs and DCs in 1970, shown in Table 4.2, reveal that the average figure for the selected developing countries is well above half that shown for the DCs.

Evidently, advertising cannot be regarded merely as a product of affluence. Its absolute (and proportionate) levels in many poor countries are very significant. Almost certainly there is far more advertising expenditure in modern LDCs than there was in the now developed countries at comparable levels of affluence.[10]

The explanation

Why is advertising much more salient in the LDCs than the Galbraithian thesis would allow? Fundamentally, Galbraith is wrong in asserting that 'the need and the opportunity to persuade people arise only as people have the income to satisfy relatively unimportant wants'. Even when people are so poor that they have difficulty

Table 4.2

Countries	*1970 advertising as per cent of manufacturing product*
United States of America	8.11
Fed. Rep. of Germany	3.63
Japan	3.17
Canada	6.32
France	1.99
Italy	1.75
Australia	6.26
Spain	3.70
Turkey	4.82
Total (selected developed market-economy countries)	6.03
Brazil	3.84
Argentina	4.86
Mexico	2.97
Venezuela	2.17
Chile	3.43
Egypt	2.62
Israel	2.95
Thailand	1.38
Saudi Arabia	1.92
Guatemala	1.37
Ghana	2.72
Total (selected developing countries)	3.60

Source: D. Chudnovsky (July 1979). 'Foreign trade-marks in developing countries', *World Development*, Vol. 7(7), pp. 663–82.

in achieving minimal living standards, they nonetheless may face important consumption choices, and what they choose may be influenced by advertising. There are two aspects to this. On the one hand even given and basic wants can be met in a variety of ways.[11] On the other hand the ordering of wants may itself be altered – 'psychological' wants, to use Galbraith's terminology, are substituted for physical ones. Moreover, the argument that the needs of the poor are directly revealed to them through their physiological senses overlooks the fact that the physical well-being of many depends on purchasing decisions made on their behalf by others.

Table 4.3 India, 1976–77

Category	*Advertising as % of sales*
Cosmetics, toiletries	2.2
Hotels	2.1
Pharmaceuticals	1.9
Entertainment goods	1.6
Vegetable products	1.5
Electrical goods	1.5
Paints, varnishes	1.2
Air conditioners and refrigerators	0.7
Food products	0.5
Man-made fibres	0.4
Cotton textiles	0.4
Autotyres and rubber goods	0.4
Cigarettes	0.4
Engineering goods	0.4
Chemicals and chemical goods	0.4
Glassware	0.3
Cement	0.1

Source: 'Review of management', *Economic and Political Weekly* (26 August 1978), p. 93.

Part of Galbraith's inability to see that the non-affluent *could* be subject to advertising can be attributed to his failure to foresee the markedly changed environment of today's, compared with yesterday's LDCs, in particular the rapid growth of transnational corporations (TNCs) and the associated spread of media technology.

A major oligopolistic or monopolistic advantage of TNCs lies in their marketing expertise, of which advertising is an important component. While operating in LDC markets TNCs exploit this advantage to the full. Thus, Raymond Vernon's data for 187 American multinationals listed among the largest 500 industrial firms in 1967 showed that the former spent a higher proportion of sales on advertising than both the remainder of the 500 large producers and United States manufacturing enterprises in general.[12] Another study by Stopford and Wells of 87 multinational firms for which advertising data were available showed that the firms with relatively high percentages of total business conducted abroad also spent relatively large proportions of total sales on advertising.[13] When TNCs first launch a 'mature' product in foreign markets 'marketing is the principal competi-

Table 4.4 Percentage share of advertising time in total hours of radio and television broadcasting

Group of countries	*Per cent in radio*	*Per cent in television*
Developed market-economy countries	5.80	4.91
Socialist countries of Eastern Europe	0.89	2.20
Developing countries	19.84	11.85

Source: D. Chudnovsky (July 1979). 'Foreign trade-marks in developing countries', *World Development*, Vol. 7(7), pp. 663–82.

tive weapon available to managers of foreign subsidiaries'.[14] Not all goods however rely equally heavily on advertising expenditure to promote sales. Table 4.3 for example shows the most heavily advertised goods in the Indian case.

This result is confirmed by evidence from other LDCs which shows similar high rankings for cosmetics and pharmaceuticals.[15] Not surprisingly therefore advertising has been shown to play an especially significant role in those TNCs producing branded pharmaceutical goods, soaps, and detergents and toiletries.[16]

Accompanying, and indeed facilitating, the rapid growth of TNCs has been the vastly increased availability of advertising media in the LDCs. The data in Table 4.1, column 4, strongly suggest that TV, cinema and radio are relatively more important advertising media in LDCs than in DCs. These have the obvious advantage over print of being accessible to the illiterate and the remote. There is also evidence that a greater proportion of broadcasting time is devoted to advertising in LDCs than in DCs. Table 4.4 illustrates.

The advertising industry which employs the media in LDCs is highly internationalized. The main reason for this is the close relationship between the TNCs and international advertising agencies, many of which established branches abroad specifically to service the operations of major clients. The relationship between the multinational producers and advertising agencies is one of mutual support in which 'a changed communications structure increasingly transmits and reinforces the attitudes that fit nicely with the requirements of the multinational goods producers that are financing the new system'.[17] The international advertising industry is very highly concentrated – 21 of the largest 25 agencies in the world are American-owned or are strongly linked with US agencies. Data presented by Sauvant show that, in 44 out of the 73 countries included in the study, the largest advertising agency is foreign majority owned, and in 39 of these by American parent agencies or those with a high American participation.[18] In four countries of Latin America – Mexico, Brazil, Argentina and Venezuela – 54 per cent of the largest advertising agencies were US-owned or affiliated.[19]

Finally, we do not overlook the obvious point that the Galbraithian view, while substantially incorrect, is not wholly without foundation. Within non-affluent societies exist small minorities of wealthy individuals for whom a good deal of LDC advertising is undoubtedly intended. Obviously even advertising aimed at the rich may influence the poor both directly and via the effect it has on the behaviour of the rich. We examine this point further in part III.

The broader context

Advertising is heterogeneous and, thus, somewhat resistant to definition. Moreover it constitutes merely a part of the general cultural environment and flows of information to which people are exposed. Indeed, one of the most striking features of the poor countries today is the extent to which their cultural environment is influenced and their information flows are dominated by the DCs.

We have already alluded to the domination of world advertising by international and especially US firms. This is only part of a more general trend, however. In particular there has been a marked general trend towards concentration in world communications.[20] This can be seen in the case of films and television programmes as well as printed material.

In many cases imported films amount to 90 per cent of those exhibited. Less than 12 large production and distribution companies (known collectively as 'Hollywood'), in fact, keep half the cinemas in the non-socialist world supplied with films.[21]

The pattern is similar in relation to the American share in programming of world television. For many developing countries the import (and American) component of programming is very high. In the Commonwealth Caribbean, for example, the proportion of imported programmes in prime viewing time is 88 per cent (excluding the nightly news).[22] American influence via the radio is also marked – United States sales of records and cassettes account for 70 per cent of non-socialist country world sales.[23]

Finally, as far as printed material is concerned, it may be noted that although the share of the developing countries in the total school population was 63 per cent in 1970, their share of global book production was only 19 per cent.[24]

In short, the total flow of communication taking place between the industrialized part of the world and the LDCs is overwhelmingly towards rather than from the latter.[25] It is tempting to draw a distinction between 'ambient' and 'conscious' advertising. The pervasive influence of DC culture itself constitutes a powerful advertisement for the general and specific components of that culture and may be expected to have some of the same effects as advertising consciously directed at LDC markets.

We continue in part III with an explanation of the ways in which advertising may be expected to have effects in LDCs that are significantly different from those which typically obtain in DCs.

III

Differential effects of advertising in LDCs

The differential nature of information

While it would be naive to assert that advertising is predominantly informative, many advertisements do contain at least some information regarding the products being promoted. Some information (such as where and when sales are being held) is unambiguously helpful to consumers. Most of the information, however, is of a different kind and is concerned to draw attention to one or more of the characteristics of the advertised goods. One study has attempted to classify advertisements according to the nature of their information content. The survey was based on advertisements on (prime time) television and in three mass-circulation magazines.[26] The results are shown in Table 4.5. Most advertising stressed *particular* characteristics of products and was rarely of a comparative kind. Though this data pertains to DCs there is little reason to suppose that it would differ in the developing-country context.

There is little doubt that consumers in general can be induced to pay more attention to one product characteristic (such as colour) rather than to another (such as texture) through advertising that stresses colour and its importance.[27] By the same token consumers may be made to pay no or little attention to the presence of undesirable or harmful characteristics of products. Either way, objective

Table 4.5

	*Television** (%)	*Magazine*† (%)
1. Ads explicitly describing the overall work, value or goodness of the advocated product	14	12
2. Ads explicitly asserting that the advocated product has certain attributes or causes certain outcomes (other than a global worth-type attribute)	80	94
3. Ads explicitly asserting that the advocated product has certain attributes or causes certain outcomes which in turn cause certain outcomes	45	72
4. Ads explicitly asserting that one evaluative dimension is the most important or is relatively more important than another in evaluating products like that advocated	3	9
5. Ads making (general) explicit comparative assertion about the advocated product on a particular dimension	18	24
6. Ads explicitly positioning the advocated brand *vis-à-vis* identified competing products on a dimension and offering an assertion that this dimension should be the primary consideration	7	5
7. Ads explicitly asserting that the advocated product has several features and that any product without these is unqualified	4	1
8. Ads explicitly asserting that when all features are weighed or balanced or integrated, the advocated brand is superior	1	4
9. Ads explicitly citing negative consequences of not adopting the advocated brand	4	2
10. Ads making explicit assertions about the benefits or advantages of some choice procedure	0	0

Source: M.L. Ray and S. Ward (eds) (1976), *Communicating with Consumers*, Sage p. 68.
* n = 120.
† n = 532.

characteristics of goods are either under- or overstated to suit the aims of the marketer. That imperfect information of this kind transmitted through advertising lowers the welfare of consumers can be shown by the simple diagram shown in Figure 4.1.[28]

To begin with, there are two goods represented by *OX* and *OY*, and two characteristics, 1 and 2. The efficiency frontier or budget line is given by *HJ*. Now assume that an advertising campaign exaggerates the extent of characteristic 2 contained in good *X*. As a result the ray *OX* shifts to *OZ*. Thus, when the consumer is actually at point *H* on *OX* he thinks himself to be at *L* on *OZ*. The 'notional' or imaginary

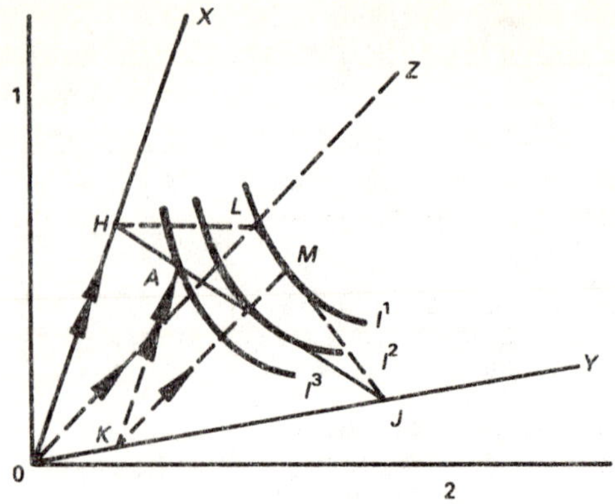

Figure 4.1

efficiency frontier becomes *LJ* and the consumer chooses his optimal position at *M* on it. In this position he wants to consume *OK* of *Y* but given this expenditure and unchanged relative prices he can then only reach *A* on the actual frontier *HJ*. Here he is on a lower indifference curve (I^3) than he would have been (I^2) given full information.

Given the welfare impact of misinformation, the question we are concerned with is whether there is any tendency for advertising to exaggerate positive and/or to understate negative product characteristics to a greater extent in developing than in DCs. In fact, there is evidence for both of these tendencies.

There is now growing documentation of advertising by the drug, food and tobacco industries which conveys a greater degree of misinformation than that provided in DCs.[29] The examples are too numerous to enumerate here but it seems clear that the practice is sufficiently widespread to warrant an examination of its underlying causal factors.

Perhaps the most obvious reason is that developing countries as a whole lack standards of advertising practice or at least have standards which are low relative to DCs. In recent years, in response to powerful consumer groups, important disclosure requirements have been introduced in the DCs. In the USA, for example, these include nutrition labelling, octane labelling, phosphate content of detergents, open dating of foods, cigarette health hazards, etc.[30] Yet in the developing world perhaps the majority of countries have a total absence of consumer protection laws.[31] In India, for example, the lack of regulation of advertising has been deplored in a recent review of management where it is argued that 'what is required of advertising in India is a code of standards whose implementation is supervised by a statutory authority. Similarly, false and misleading claims, if they are made, must be punishable offences.'[32]

The situation of many of those living in developing countries can thus be likened in one way to that in the industrialized nations in the 1920s and 1930s when consumers were 'purchasing new and unfamiliar consumer durables and a growing array of other new products with little information to go on except that gleaned from the deluge of advertising to which they were subjected'.[33] In fact the situation is even more difficult in the modern developing countries in so far as products have grown in complexity and are less amenable to comprehension by consumers without much education.

The differential perception of information

Not only, as we have noted, does the information conveyed by advertising tend to differ for those in developing countries but it is also likely to be differently perceived.

In the DCs advertising operates within a particular ethos. That is, the consuming public accord business in general, and advertising in particular, a low rating for integrity. In a nationwide survey in the USA in 1972, 60 per cent of consumers sampled felt that recent criticisms of advertising were totally justified while less than half of all advertisements were rated as honest and informative.[34] As a result of these attitudes, it is not surprising that many advertising claims are heavily discounted.[35] Moreover, a study which attempted to measure consumers' propensity to be deceived by the false claims of advertisers found that, 'Younger, less educated, lower income and lower status occupation consumers exhibited a greater tendency to be deceived.'[36] By extension one would expect this finding to apply to many, if not most, of those living in developing countries – only more so. These consumers have had insufficient time to develop the scepticism which as noted above has evolved on the part of those living in rich countries. In fact many in the Third World tend to regard those in rich countries as 'authority figures' so that advertising claims made on behalf of goods from these countries are at once invested with a high degree of credibility. There is unfortunately very little evidence concerning the comparative propensities for deception in rich and poor countries but the few pieces of information available for LDCs are supportive of the likelihood of marked differences.[37]

Advertising within different competitive structures

We have already referred to Galbraith's view that in low-income societies 'all the commercial advantages' lie with the producers of simple goods (plain bread, sidemeat and oatmeal) as opposed to those producing sophisticated varieties of the same goods (sliced and wrapped bread, canned ham and processed breakfast foods). In fact, this is by no means generally true in developing countries.

The size and ownership structures of firms in developing countries are very different from those in the developed world. In particular there are very many more firms with only a small number of employees, many of which operate outside the wage–labour system. Far from possessing all the commercial advantages, these firms are often distinctly at a disadvantage when confronted by competing multinational subsidiaries. We noted above Vernon's finding that American multinationals spend a higher proportion of sales on advertising than other industrialists. There is also some evidence to support the claim that multinationals devote more expenditure to advertising than competing local firms in developing countries. Thus, for a variety of manufacturing industries in Malaysia, Chong concluded that on average foreign firms spent some 6.2 per cent of sales on advertising compared with the figure of 1.6 per cent for local firms.[38] Langdon reached similar conclusions in his study of soap manufacture in Kenya.[39] In the same country it is reported that among the 10 firms with the largest absolute expenditures on advertising in 1975 there were no locally owned firms.[40]

Therefore, while competition in modern developed societies takes place mainly on the basis of advertising and product differentiation between very large oligopolists, this is not the typical situation in developing countries. Through heavy advertising, multinationals in the latter are relatively easily able to secure market predominance for their particular branded goods.[41] An interesting manifestation of this difference emerged from a study in which a sample of 400 Kenyan and 400 English consumers were requested to identify four products which were shown to them merely as line drawings with no clues as to brands. The result was that 50 per cent of the Kenyan sample described the four goods (toothpaste, cigarettes, aspirin tablets and a razor blade) in terms of their brand names as opposed to the 2 per cent of the English sample that did so.[42]

Advertising as a high-income characteristic

The history of product developments reveals two related tendencies. First, new products developed almost exclusively in and for DC markets embody an increasingly large proportion of high-income characteristics to match the rising incomes in these markets.[43] Accompanying this trend has been a rise in the proportion of selling costs to total costs of production. Ever increasing sophistication and complexity of goods, that is to say, require as an essential concomitant an increasing emphasis on selling costs of which advertising is the major component. The problem is that 'the progressive substitution of non-price for price competition tends increasingly to offer consumers a choice from among competing packages of goods, with no possibility to buy only part of the package and reject the rest'.[44] From the point of view of those with very low incomes this indivisibility is clearly more marked than it is for higher-income consumers.[45]

In other ways, too, advertising is likely to have an inegalitarian impact in LDCs. To the extent that it causes a switch from local to foreign brands, the result is likely to favour foreign owners at the expense of domestic employment and poor consumers, who are forced to consume the more expensive foreign brand.[46]

Advertising may also have an impact on the intergenerational distribution of incomes in LDCs. If it raises the aggregate marginal propensity to consume, thus reducing savings and investment, advertising will exert a constraint on the growth rate and, hence on the interests of future generations.

IV. CONCLUSIONS

We have shown that, contrary to J.K. Galbraith's view, advertising is not merely a product of affluent societies and that for many poor countries it is indeed a significant phenomenon. Yet, as we showed in part III, the effects of advertising in poor countries are likely to differ sharply in a number of important areas from those in DCs. In none of the areas, however, are data sufficient and there is a clear need for further research.

At no stage in the chapter have we referred to the impact of advertising on taste formation in LDCs. This is not because we share the traditional disdain of economists for the subject, or because we think it unimportant. The problem is rather a notable lack of hard evidence on the impact of advertising. We believe that detailed research is required in order to establish how tastes are formed in LDCs. Such research might confirm what some have long suspected: that economists' traditional 'neutrality' towards consumers' tastes, far from being value-neutral, involves the acceptance of tastes which are: (1) based on a greater degree of misinformation than those in DCs; (2) based on a highly unequal quantity of advertising between competing varieties of goods meeting the same broad needs; (3) formed by consumers who are seldom able to process adequately the information conveyed to them by advertising.

Notes

1. For a brief statement of the view that advertising benefits LDCs, see Stridsberg (1971).
2. Galbraith (1973), p. 150.
3. Ibid., p. 152, emphasis added.
4. Galbraith (1963), p. 114, emphasis added.
5. Ibid., p. 111.
6. Ibid., p. 114.
7. Ibid.
8. Galbraith comes very close to stating (2) explicitly when he argues that, 'The further a man is removed from physical need the more open he is to persuasion – or management – as to what he buys' (Galbraith, 1969, p. 207).
9. The regression equation is:

$$\underset{}{y} = \underset{(5.14)}{0.01x} + \underset{(6.04)}{41}, \qquad \bar{R}^2 = 0.31,$$

   where y is advertising as a percentage of national product and the independent variable is national product.
10. Roughly 100 years ago advertising scarcely existed in the USA. See Baran and Sweezy (1970).
11. In this case a necessary condition is that goods be advertisable in the sense that their true properties are obscure to a greater or lesser extent. For an attempt to differentiate between products on this basis, see Greer (1979).
12. Vernon (1971).
13. Stopford and Wells (1972).
14. Ibid., pp. 55–6.
15. In Colombia over 50 per cent of television advertisements are devoted to cosmetics, non-essential foodstuffs and detergents, most of which are

produced by US multinationals. See Bettran (1977). For Argentina and Mexico see, respectively, Chudnovsky (1979) and Greer (1979).
16. See G.K. Helleiner (1975). Stopford and Wells also found that multinationals in the food, beverages, detergent and drug industries had the largest expenditures on advertising. These firms, with particular expertise in marketing, were very reluctant to have anything other than wholly-owned foreign subsidiaries. Two factors were held to be responsible for this orientation: namely, the availability of marketing skills within the firms and the commitment to a given marketing strategy. See Stopford and Wells (1972).
17. Quoted in Sauvant (1976), p. 58.
18. Ibid.
19. Barnet and Muller (1974).
20. Varis (1976).
21. Ibid.
22. Hosein (1976).
23. Varis (1976).
24. Smith (1977).
25. Varis (1976).
26. Wright and Barbour (1976).
27. Stefflre (1968).
28. Auld (1972).
29. Medawar (1979).
30. Day (1978).
31. Medawar (1979).
32. 'Review of Management', *Economic and Political Weekly*, 26 August 1978, p. 69.
33. Herrman (1978), p. 29.
34. Wilkes and Wilcox (1978), p. 223.
35. For some evidence see Armstrong, Kendall and Russ (1976).
36. Ibid., p. 48.
37. A study of food purchasing in Calcutta, for example, found that 'in the lowest income groups, 61 per cent rated branded baby foods "good for the health of the child" against 41 per cent in the highest income group.' Quoted in Medawar (1979), p. 104. In Africa the average consumer has been described as 'susceptible, in fact, very receptive to advertising' (ibid., p. 38).
38. Quoted in Medawar (1979).
39. Langdon (1975).
40. Medawar (1979).
41. Thus in a developing country the marketer can 'economically create an almost unassailable position of strength in the market place, which would be unattainable in a more sophisticated environment.' Quoted in Medawar (1979), p. 32.
42. Ibid.
43. Stewart (1977).
44. Scitovsky (1964).
45. James and Stewart (1981).
46. Langdon's study illustrates this particular scenario. See Langdon (1979).

References

Armstrong, G.M., Kendall, C.L. and Russ, F.A. (1976). 'Application of Consumer Information Processing Research to Public Policy Issues', in Ray, M.L. and Ward, S. (eds), *Communicating With Consumers*, Sage.

Auld, D.A.L. (1972). 'Imperfect Knowledge and the New Theory of Demand', *Journal of Political Economy*, Vol. LXXX, Nov.–Dec.

Baran, P.A. and Sweezy, P.M. (1970). *Monopoly Capital*, Pelican.

Barnet, R.J. and Muller, R.E. (1974). *Global Reach*, Simon & Schuster.

Bettran, L.R. (1977). 'TV Etchings on the Minds of Latin Americans: Conservatism, Conformism and Other Patterns', *International Development Review*, Vol. XIX (X).

Chudnovsky, D. (1979). 'Foreign Trade-Marks in Developing Countries', *World Development*, Vol. 7 (7), July.

Day, G.S. (1978), 'Assessing the Effects of Disclosure Requirements', in Aaker, D.A. and Day, G.S. (eds), *Consumerism*, Free Press.

Galbraith, J.K. (1963). *American Capitalism*, Pelican.

Galbraith, J.K. (1969). *The New Industrial State*, Pelican.

Galbraith, J.K. (1973). *The Affluent Society*, Pelican.

Greer, D. (1979). 'The Economic Costs and Benefits of Trademarks: Lessons for the Developing Countries', *World Development*, Vol. 7 (7), July.

Helleiner, G.K. (1975). 'The Role of Multinational Corporations in the Less Developed Countries' Trade in Technology', *World Development*, Vol. 3 (4), April.

Herrman, R.O. (1978). 'The Consumer Movement in Historical Perspective', in Aaker, D.A. and Day, G.S. (eds), *Consumerism*, Free Press.

Hosein, E.N. (1976). 'The Problem of Imported Television Content in the Commonwealth Caribbean', *Caribbean Quarterly*, Vol. 22 (4), Dec.

James, J. and Stewart, F. (1981). 'New Products: A Discussion of the Welfare Effects of the Introduction of New Products in Developing Countries', *Oxford Economic Papers*, Vol. 33 (1). Reprinted as Chapter 2 above.

Langdon, S. (1975). 'Multinational Corporations' Taste Transfer and Underdevelopment', *Review of African Political Economy*, Jan.–April.

Medawar, C. (1979). *Insult or Injury? An Enquiry into the Marketing and Advertising of British Foods and Drug Products in the Third World*, Social Audit, London.

Sauvant, K.P. (1976). 'The Potential of Multinational Enterprises as Vehicles for the Transmission of Business Culture', in Sauvant, K.P. and Lavipour, F.O. (eds), *Controlling Multinational Enterprises: Problems, Strategies, Counterstrategies*, Westview.

Scitovsky, T. (1964). *Papers on Welfare and Growth*, Allen & Unwin.

Smith, K.B. (1977). *The Impact of Transnational Book Publishing on Intellectual Knowledge in Less Developed Countries*, UNESCO, Division for the Study of Development.

Stefflre, V. (1968). 'Market Structure Studies: New Products for Old Markets and New Markets (Foreign) for Old Products', in Bass, F.M. *et al.* (eds), *Application of the Sciences in Marketing*, Wiley.

Stewart, F. (1977). *Technology and Underdevelopment*, Macmillan.

Stopford, J.M. and Wells, L.T. (1972). *Managing the Multinational Enterprise*, Basic Books.
Stridsberg, A. (1974). 'Can Advertising Benefit Developing Countries?', *Business and Society*, Autumn.
Varis, T. (1976). 'The Impact of Transnational Corporations on Communication', *International Social Science Journal*, Vol. XXVIII, No. 4.
Vernon, R. (1971). *Sovereignty at Bay*, Basic Books.
Wilkes, R.E. and Wilcox, J.B. (1978). 'Recent FTC Actions: Implications for the Advertising Strategist', in Aaker, D.A. and Day, G.S. (eds), *Consumerism*, Free Press.
Wright, P. and Barbour, F. (1976). 'The Relevance of Decision Process Models in Structuring Persuasive Messages', in Ray, M.L. and Ward, S. (eds), *Communicating With Consumers*, Sage.

Appendix
Some Recent Findings

The findings based on Table 4.1 tend to be confirmed by studies using more recent data sets. Jenkins, for example, reports evidence for the late 1970s that show very high world rankings for several developing countries based on the ratio of advertising to GDP. More generally, his data reveal 'little evidence of a clear relationship between level of development and advertising. Thus Jamaica has similar levels to Britain; Singapore, Thailand and Venezuela rank above France and Colombia and South Korea ahead of West Germany' (Jenkins, 1988, p. 1365). Data for the 1980s cited by Austin show that 'Advertising levels increase proportionally less than per capita income levels' (Austin, 1990, p. 282).

References

Austin, J. (1990) *Managing in Developing Countries*, Free Press.
Jenkins, R. (1988), 'Transnational Corporation and Third World Consumption: Implications of Competitive Strategies', *World Development*, Vol. 16 (11).

5 The Developed-Country Impact on Product Standards in the Third World

INTRODUCTION

Legal minimum standards below which products may not be marketed are widely used in developed countries. There is a commonly held presumption that these standards – almost all of which are imposed for goods endangering the health and safety of consumers – should also be adopted by the developing countries. In both groups of countries, for example, there is legislation prohibiting the marketing of drugs which are not approved in the United States. The World Health Organization formulates international standards for drinking water and drugs,[1] while consumer groups urge 'that the United States support the development of international product safety standards'[2] and express indignation at the very idea that drugs which do not meet developed-country standards should be available in developing countries.[3]

Many governments in developing countries have unquestioningly adopted developed-country product standards in sectors such as construction, transport and health.

The purpose of this chapter is to document the transfer of product standards from developed countries and to assess the appropriateness of these standards for those living in the developing countries. To begin with, it is useful to consider why standards are adopted in the developed countries and thereafter to understand why they are set at particular levels.

THE BASIS FOR INTERVENTION THROUGH STANDARDS IN DEVELOPED COUNTRIES.

A fundamental assumption of Paretian welfare theory is that 'an individual should be considered the best judge of his economic welfare'.[4] This value judgment is often referred to as that of 'complete consumer sovereignty' and its rejection is perhaps the most basic reason for the adoption of standards in developed countries.

Paternalism

The need for government protection of particular groups in society – such as children and the insane – who are obviously unable adequately to assess their own interests has long been recognized. Some product standards, such as seat-belts for children, reflect this need for paternalism. Clearly, however, different reasons have to be adduced to explain why individual freedom of choice is abrogated in the case of standards (such as those relating to compulsory seat-belts and crash helmets) which are imposed on the *entire* population. One possible justification is the view that individuals tend systematically to underestimate the true risks of injury even when they are aware of these risks. It is well-known from research in social psychology, for example, that individuals often use techniques of rationalization (including the distortion of available information) to justify their own behaviour (smokers, for example, are known to underestimate the dangers involved in smoking compared to non-smokers).[5]

A similar type of argument, which may underlie some of the safety standards on housing, is that individuals need to be protected from harming themselves because they are typically short-sighted about risks occurring in the future – a form of Pigou's 'defective telescopic faculty'.

More generally, there is a large literature on the psychology of decision-making under risk, much of which questions the rationality of the way that we commonly deal with the problem.[6]

Externalities in consumption

In some cases standards which seem to be based on paternalism may also (or alternatively) be due to externalities in consumption, i.e. the effect of one individual's consumption on the welfare of others. The reason is that many of the risks undertaken by individuals are borne also by society. For 'even a person who harms only himself will often have to be rescued, patched up, nursed back to life, or buried with the aid of public services and partly or wholly at taxpayers' expense'.[7]

And to protect the taxpayer, governments may restrict the freedom of the individual to decide what chances to take with his own safety. Legislation requiring that motorcyclists wear crash helmets, for example, may be due partly to paternalism and partly to a view that 'reckless' individuals should not impose their preferences on the rest of society.

In the presence of externalities product standards may also be required partly to *give effect* to individual preferences. As an illustration consider the question of speed limits and assume a situation in which for each individual in society the most desirable situation is one in which everyone except himself has to observe a speed limit. Provided that everyone else does observe the limit, however, he would prefer to have to adhere to it rather than have no speed restrictions. In this sort of situation, known in game theory as the isolation paradox, government intervention to enforce the speed limit is required not to overrule individual preferences, but rather to give effect to them through securing the collectively rational outcome. A similar case could be made for gun control, despite the fact that on revealed preference grounds, many Americans seem to prefer to possess guns. The case is this. In a situation of widespread violence:

> an individual may arm himself for his self-protection, and his choice then suggests that he prefers owning a gun to not owning a gun. If all people make the same decision, then their choices seem to indicate that they prefer not to have ownership of guns outlawed. In fact, however, there may be crucial unrevealed preferences. Each individual may prefer (1) owning a gun but having it outlawed for all others, to (2) having it outlawed for everyone, to (3) everyone owning a gun, to (4) not owning a gun while all others own one. Whether the others own guns or not, individuals prefer to own guns; they prefer (1) to (2) and also (3) to (4). The market in this case tends towards outcome (3). Yet all the individuals prefer (2), that is, a general prohibition of guns, to (3).[8]

Imperfect information

The paternalistic and externalities arguments for standards apply even when it is assumed that the consumer is fully informed. When this assumption is dropped, an additional case for standards sometimes arises.

It is difficult to deal adequately with imperfect information in terms of the traditional theory of demand which assumes that goods as such are the immediate object of utility. The reason is that the scope for imperfect information about goods in this theory is limited to that about their existence and prices. The characteristics approach, in contrast, which as we have seen views the demand for goods as a derived demand for the characteristics embodied in them, permits us to analyse a situation in which the consumer is imperfectly informed about any or all the relevant characteristics of a good. (See Chapters 2 and 4.)

The fact that buyers are imperfectly informed in this sense does not of itself, however, indicate the need for product standards to be imposed by government. The 'first-best' solution to the problem of imperfect information is, of course, the provision of full information about product characteristics because it preserves freedom of choice. A zealous advocate of this approach to consumer protection is Milton Friedman, who, together with Rose Friedman, argues that 'Insofar as the government has information not generally available about the merits or demerits of the items we ingest or the activities we engage in, let it give us the information. But let it leave us free to choose what chances we want to take with our own lives.'[9] The requirement that a health warning and statement of tar content be carried on cigarette packets and the compulsory labelling of many other categories of potentially dangerous products (such as flammable textiles and household appliances) in most developed countries indicate that the market mechanism, aided by the provision of information, is often thought to be superior to direct regulation through the imposition of standards.

In some cases, however, the provision of information is unlikely to contribute to more informed choices by consumers. Technical information about the safety of drugs, for example, is unlikely to make much sense to the layman. Moreover, even if complex information can somehow be presented in a comprehensible way to individuals, the costs of doing so would often appear to be prohibitive. Where, for these reasons, information is difficult or impossible to communicate to consumers, reliance on the market mechanism becomes more difficult and product standards may be required to protect uninformed consumers. What has to be decided is whether the benefits of these consumers through the protection thus afforded by standards are sufficient to outweigh the costs imposed by the restriction on the range of products available. Following Colantoni *et al.*,[10] the problem may be analysed with the aid of Figure 5.1.

Figure 5.1 portrays a situation of two goods *X* and *Y*. The prices of these goods are such that the slope of the efficiency frontier is given by *BD*. Two individuals, with the same incomes but different tastes, are represented by the indifference maps *ii* and *II*. Possessed of perfect information the former would choose good *X* and the latter good *Y*. Let us assume, however, that the individual with the *ii* map overestimates the characteristics embodied in good *Y* to the extent

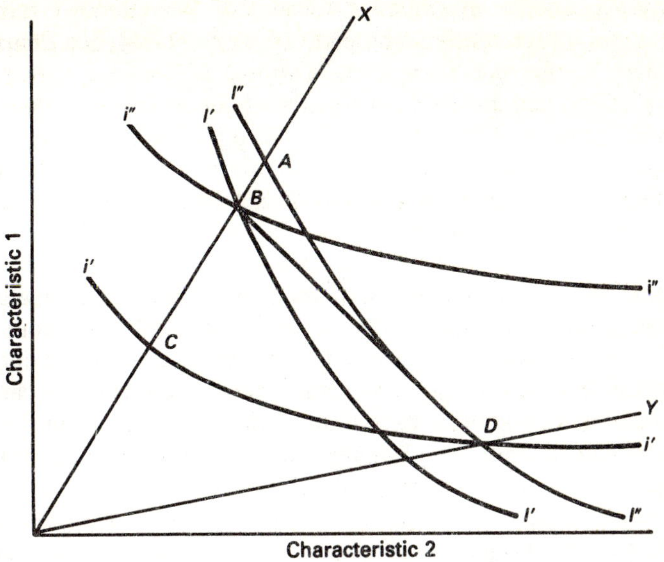

Figure 5.1 The welfare effect of product standards.

that he chooses it in preference to good *X*. The effect of government intervention to remove good *Y* then implies that the individual with the *ii* map is made better off (because he is saved from making a mistaken choice) and the individual with the *II* map is made worse off, because his optimal choice is no longer available. In terms of good *X* the gainer could in this case compensate the loser and still remain better off, i.e. $BC > AB$. The effect of the withdrawal of *Y* on the distribution of income depends partly, of course, on whether or not compensation is actually paid to the loser.

In any actual situation, therefore, the argument in favour of the withdrawal of a particular good will be stronger:

1. the greater the extent to which individuals are misinformed about the characteristics embodied in it (or more specifically, overestimate its positive characteristics and underestimate its negative characteristics) and the more serious are the risks associated with its consumption;
2. the fewer are those for whom the good is the optimal choice;
3. the more the change leads to a desired movement in income distribution.

While, of course, it cannot be said that governments actually calculate the relative size of these effects, it can be argued that they underlie in a rough way many of the decisions to ban dangerous drugs and chemicals that are made in developed countries.

DEVELOPING COUNTRY REQUIREMENTS FOR PRODUCT STANDARDS

The case for *some* standards in developing countries rests on the same grounds as those described above in relation to developed countries; namely, that because of paternalism, externalities in consumption and imperfect information, the unregulated market mechanism produces insufficient protection of consumers. But it does not follow from this that countries of the Third World should adopt either the same *number* or the same *level* of standards as the developed countries.

Where imperfect information is the problem underlying the need for product standards, it is understandable that in the case of many goods (some of which were described above), policy-makers in developed countries prefer modification of the market mechanism through the provision of information to direct regulation through standards. In these societies high levels of education and efficient communications networks enable much information to be provided and assimilated by consumers relatively easily. The situation in most countries of the Third World, however, is very different. Low overall rates of literacy (particularly in Africa, for example) and educational attainment, and limited coverage of the mass media (especially in rural areas), make it much more difficult and sometimes impossible to rely on the provision of information. The experience of nutrition and other mass education programmes in developing countries illustrates the difficulties of successfully conveying information to isolated and dispersed rural communities, many of whose members are illiterate. As a result of these difficulties, the role of standards in protecting uninformed consumers may need to be greater. But as we saw above, for the imposition of standards to be justified, the benefits afforded through protection have to exceed the costs incurred through the restriction of product choice. This, in turn, depends on the *level* at which the standards are set.

DEVELOPED-COUNTRY LEVELS

Income is an important determinant of individual tastes, as previous chapters have indicated. Even if Maslow's notion of hierarchical needs is subject to important limitations, it is nevertheless true that at the very low income levels characteristic of the majority of those living in developing countries, products are demanded to a substantial degree for basic needs that they help to meet.

It is also true that as societies become richer they require increasing standards of safety from the goods that they purchase. New products developed in the rich industrialized countries often reflect this relationship. Cars and jets, for example, have both become safer over the years. In the case of drugs, 'The current trend of society in demanding quantitative increases in the quality of life has programmed itself, in terms of ethical pharmaceuticals, to accepting a lower level of risk.'[11] This is undoubtedly a major reason for the fact that 'the consumption of pharmaceuticals is now probably safer than at any time in pharmaceutical history.'[12] Indeed, a number of highly valued drugs in common use, such as penicillin and aspirin, would more than likely have failed toxicity tests by the current standards of a number of DCs.[13]

Product safety standards in these countries, such as those administered by the FDA (Food and Drug Administration) and the CPSC (Consumer Product Safety Commission) in America, reflect in a rough way the increased premium placed by these societies on safety as incomes rise. At the same time they are designed to fit in with the skills and other resources now possessed by these countries. For example, manually-operated X-ray machines are no longer approved in the USA because of the risk to patients of unnecessary exposure. But the introduction of complex machines that automatically limit the beams to the area to be X-rayed was made possible by the availability of technical skills needed to operate the new equipment.

The safety of many new goods developed in and for DCs is not, however, achieved without a cost. Increases in cost (and price) take the form not only of additional characteristics of goods (such as seat-belts, padded instrument panels and energy-absorbing steering columns) but also more extensive product testing. Some early antibiotics, for example, were developed, tested and approved within eighteen months to two years. Now a period of five to seven years is considered normal.[14] At the same time the cost of testing new chemical entities has increased dramatically.[15]

It is very likely that the individual consumer in the typical LDC will have a view of the extent to which additional safety should be 'bought', which is generally quite different from that of his counterpart in the DCs. A basic reason for the difference lies in the close relationship (noted above) between incomes and some basic human needs. By way of an (all too common) example consider the situation of a head of a low-income household in an LDC who is taken ill and is unable to work. Deprived of any form of sickness benefit he or she is aware that even a relatively short period of loss of earnings can mean severe physiological hardship for himself/herself and his (or her) family. In this situation he or she will almost certainly have a marginal valuation of the safety, efficacy and price of drugs which differs sharply from that implied by DC standards.[16]

Empirical evidence for our view of safety as a 'high-income' characteristic is difficult to find, but one experiment conducted by Jones-Lee concerning the safety records of two airlines offers some support.[17] Each of 31 respondents was presented with the fare and safety record of a given (hypothetical) airline. They were then asked to indicate the fare that would just induce them to travel with another airline which differs from the other only in terms of safety (measured in terms of fatal crashes per 500 000 flights).

When the sample of responses is divided into two income groups (above and below an annual salary of £3500) the results support the view that safety is a characteristic whose value does seem to vary with incomes. In particular, the 'high-income' group would, on average, pay more for a 'safer' ticket; conversely, they would require greater compensation in terms of a reduction in the fare before being prepared to travel on the more 'risky flight' (i.e. they are prepared to pay more than the 'low-income' group to secure a reduction in risk on the one hand, and require greater monetary compensation to accept higher risk on the other).

THE TRANSFER OF DEVELOPED-COUNTRY STANDARDS TO THE THIRD WORLD: THE COLONIAL LEGACY

At the outset of this chapter we referred to some of the ways in which developed-country standards are transferred to the poor countries of the Third World. There are, however, a number of other ways in which this can occur, perhaps the most important of which originates in the product standards that were introduced in developing countries during the colonial period.

The British in Africa, for example, 'brought their building by-laws with them, disdaining local materials (mud, timber, thatch) more suited to the continent's heat'.[18] In more than a few African countries the application of these laws and standards appears to have extended into the post-colonial period as well. In Tanzania, for example, a 1975 Tender Document for an oil mill contains the requirement that 'Building and construction work must meet local building and hygiene requirements and standards, *which to a great extent are of British origin*.'[19] Or, to take another example from the same document, the use of locally produced roofing tiles is precluded by the 'minimum requirements for roofing materials – corrugated aluminium sheets or corrugated asbestos sheets'.[20]

In Kenya, another former British colony, 'The Undugu Society, a charity working in Nairobi's slums, has been upgrading simple homes at a cost of 8,000 Kenyan shillings [$340] each. That is illegal, because the colonial building standards enshrined in the city's by-laws make it impossible to build a house for less than ten times that figure. Even if some fairy godmother were to offer KSh80,000 for each of the city's 100,000 odd slum houses, legal slum upgrading would still be all but impossible, because the laws insist that every dwelling in Nairobi be accessible by car.'[21]

The persistence of standards such as these is perhaps most importantly a reflection of the fact that they tend to benefit the elite of post-colonial African countries, just as they did the colonial administrators before them. The housing and building standards that were just described, for example, enable the rich to distance themselves geographically from the living conditions of the poor (whose inability to meet the standards required by the city authorities is used to justify the establishment of squatter camps and townships in remote areas). 'Like the colonists before them, Africa's black elites seem to wish themselves in Europe and do not want other kinds of Africans spoiling the view.'[22]

Paradoxically, though, the reliance on developed-country product standards seems to result partly from a reaction *against* certain aspects of the colonial legacy. Consider again the case of Kenya and in particular the implementation of public housing schemes in that country over the period from 1963 to 1973, as described by Temple and Temple.[23] According to these authors, 'The question of housing standards, particularly in Nairobi, is a sensitive political issue, and in a country very concerned with its progress toward modernity, public housing is often treated as a visible symbol of Kenya's achievements. President Kenyatta himself took the lead in touting the virtues of modern housing and castigating *those who would build dwellings reminiscent of the quarters provided by the colonialists for Africans*.'[24] The President's views appeared to be shared also by other important decision-makers, whose combined opinions contributed to a general climate of 'elite opinion' that:

> . . . discouraged the kind of technical and social experimentation that might have helped surmount the obstacles to housing poorer sections of the population posed by rising building costs. When, for example, timber was used in an experimental project, the object was not to build the least expensive units possible, but to see whether it was an appropriate material for medium-cost houses. The [Nairobi city] councillor's general attitude towards the houses built in the experimental scheme was that they were nice, but would have been better if they had been built of more expensive concrete blocks.[25]

THE WELFARE CONSEQUENCES OF THE TRANSFER OF DEVELOPED-COUNTRY PRODUCT STANDARDS

In the previous section it was suggested that the rich may tend to benefit from the application of certain developed-country standards. The poor on the other hand will tend to suffer from the enforcement of these standards for they would then be denied the opportunity to purchase the cheaper (albeit often less safe) varieties that best accord with their tastes and would be forced to purchase the relatively expensive brands that conform to the standard of those living in rich countries. Figure 5.2 illustrates the argument.

AB is the efficiency frontier for the 'average' consumer in the developing country, whose tastes are represented by the indifference curves IC_1 and IC_2. The effect of the imposition of developed-country standards is to render his preferred good *Y* unavailable, leaving him on the lower indifference curve IC_2. Consequently, the welfare effect of the change is that the minority of individuals with incomes and tastes similar to those prevailing in the developed countries gain at the expense of the majority. (Note that this conclusion is not altered

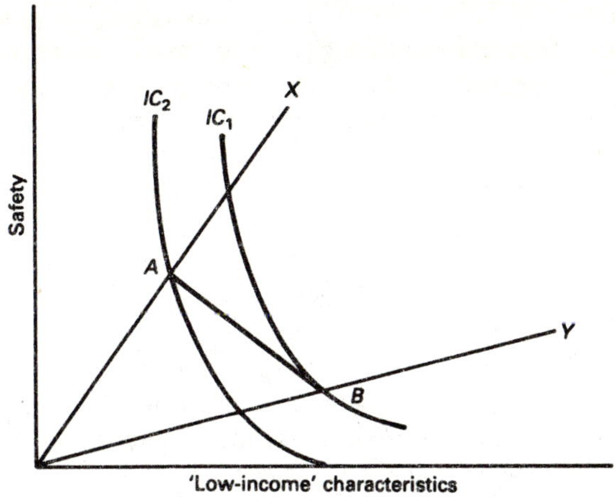

Figure 5.2 The effect of adopting developed-country standards in developing countries.

if good *Y* is bought on the basis of imperfect information; because of the assumed nature of the relationship between preferences and incomes this good would remain the optimal choice for, and continue to be purchased by, most consumers even when they possess full information.)

The examples from Africa cited earlier in connection with the colonial influence on building and housing standards well illustrate the type of situation depicted in Figure 5.2. These examples are not untypical of the Third World in general; on the contrary: 'It is common for public agencies to build houses or flats to standards which the majority cannot afford, nor can the country possibly subsidize them on a large scale. On top of this, it is not unusual for governments to prohibit private building of the type of housing the vast majority can afford and are satisfied with.'[26] Similarly, a very recent study concludes that 'In most cities, building codes will have to be revised; the standards they demand are unaffordable for the poor majority. At present, codes often deem illegal the only building materials that lower-income groups can afford, for instance mud or mud bricks, plastic sheeting and the cheapest wood based panels.'[27] Purchase of the housing with 'too-high' standards involves a disproportionate expenditure on this item compared to other items in the household budget, i.e. it leads to imbalances in consumption (or to concentration of public expenditure on housing to the exclusion of other social needs). Alternatively, because unrealistically high standards are difficult to enforce, black markets may frequently develop. In Kenya, for example, cheap maize beer brewed by low-income households and served in oil cans is banned.[28] The same restrictions apply to the sale of gin distilled in the formal sector of that country.[29] As a result, the consumers of these popular products are forced to purchase them illegally.

Still another Kenyan example has to do with the consequences for the poor of the application of housing and building standards from the colonial period that were referred to above. In particular, the illegality of the existing slum dwellings in Nairobi (which, as also noted above, could only be made to conform to the required standards by relatively large expenditures) means that their inhabitants are deprived of security of tenure. And this in turn means that 'they find it harder to borrow money to start businesses, and have less incentive to improve their homes themselves.'[30]

On occasion, the consequence of attempting to enforce unrealistically high standards in developing countries is that some consumers become even *less*-well protected. This paradoxical outcome has been observed in relation to standards for drinking water by Feachem who cites cases where health officials have closed down village tubewells because they were found to contain impurities slightly in excess of developed-country standards. As a result, the villagers were forced to use irrigation canals that were far more polluted.[31]

Mortality rates, life expectancies and discount rates

High mortality rates and low life expectancies are common features of most LDCs. Like low incomes, they may also influence the way in which safety is evaluated in these countries relative to those that are more affluent.

High maternal mortality rates in LDCs may change the risk-benefit ratio associated with contraceptives because of the higher risks of an unwanted pregnancy[32] – a factor not, of course, taken into account in FDA decisions to ban or approve new contraceptive drugs for sale on the US market. (Depo-Provera, a contraceptive administered by injection has, for example, been banned.)

Moreover, in most LDCs average life expectancies are considerably lower than in DCs. On account of the differential probability of surviving another year in DCs and LDCs Eckstein has argued that time preference – the discounting of future benefits relative to the present – will also vary.[33] In particular, the greater uncertainty of survival for individuals in LDCs implies that they discount future benefits at a higher rate than members of DCs (i.e. they weight the present more heavily). The converse presumably applies to costs which are distributed over time. As a basis for different product standards this argument is weakened by the fact that life expectancy differences between rich and poor countries are most pronounced at birth – after just one year the gap narrows markedly. Still, it is often difficult to argue that long-term risks (of, say, drugs) should be discounted in the same way in Chad (where life expectancy at the age of 20 is around 50) as in America (where life expectancy at the same age is at least 70).

Differential incidence and severity of diseases

Many diseases occur with varying degrees of severity between DCs and LDCs. Almost always, the severity is more pronounced in the latter. Measles, for example, is very rarely a lethal disease in the affluent societies. It is most severe in infancy when combined with malnutrition and other diseases endemic to the poorest segments of the Third World. Since the benefits from treatment are on this account potentially greater in LDCs, reasonable decisions to ban the use of certain measles vaccines in DCs (such as taken in the UK with the Beckenham-31 strain twenty years ago) may make little sense for many LDCs.

The incidence of diseases also varies widely as between rich and poor countries. Many tropical diseases, for example, are virtually non-existent in DCs. Though this fact will not affect the risk-benefit ratio for drugs designed to treat such diseases in the case of a single patient, it does affect the net benefits calculated for society as a whole. Conversely, some diseases in DCs, such as those of the heart, are comparatively rare in the LDCs.

THE NEED FOR INTERMEDIATE STANDARDS

In each of the areas that have just been described there is a need to adopt lower standards than those prevailing in the developed world.

What is needed may in fact be described as *intermediate* standards, in the same sense and for the same basic reason as that which underlies the widespread advocacy of intermediate technology in the Third World. This should not be taken to imply a concern only for 'downgrading' developed country standards but also for 'upgrading' local norms where necessary. For example, drinking water from surface sources or open wells in developing countries often exceeds 1000 faecal coliforms per decilitre. Raising the standard to, say, 500 faecal coliforms per decilitre is a long way from the WHO standards (of zero coliforms) but is nonetheless a significant improvement.[34]

Recognition of the need for intermediate product standards does exist but mostly in rather fragmentary form. For example, manually operated X-ray machines are no longer approved in the United States because of the risk to patients of unnecessary exposure (complex machines that automatically limit the beams to the area to be X-rayed are used instead). For the developing countries, however, 'the benefits of having X-ray equipment available for use outweighs the potential risk of unnecessary exposure from manual operation.'[35] Similar considerations apply to the administration of oral contraceptives in LDCs. Given the shortage of medical manpower in these countries, it is arguably unreasonable to require that every woman be given all the warnings, or that she get a routine smear.[36] Though this is a less-safe procedure than that undertaken in developed societies, it is surely preferable from the point of view of safety itself, as well as from other points of view, to banning the drug or restricting it to those few who are able to have the more thorough treatment. Or again, expenditure of the government health budget on cheaper (if slightly older and perhaps more risky) drugs would allow a much higher proportion of the population of a developing country to enjoy the benefits of drug therapy.

In the context of the development of African cities, it has recently been argued that standards should be relaxed to 'practicable levels'. For example:

> Illegal slums are bad, but they are better than no housing at all; and they are what you will get if you do not allow tolerable but cheap housing . . .
>
> Relaxing standards would ease transport problems too. The decay of state bus services has brought forth hair-raising private pick-ups, estate cars and vans that pack their passengers on top of one another and seldom check the tyres or brakes. They are dangerous, so city governments sometimes try to ban them. Yet

> the passengers prefer dangerous transport to no transport at all. Once again, it is a question of whom high standards serve. *The authorities must accept a half-way house.*[37]

A comparison between cement production in China and India provides a further context in which the case for intermediate standards has been made. On the one hand, following the practice of the developed countries, India has rigidly adhered to a single standard for Portland cement and no cement that fails to meet this standard can be sold for general use. China, on the other hand, permits no fewer than six standards of cement quality; these allow different qualities of cement to be used for purposes that have variable demands. Relatively low qualities, for example, are said to be perfectly adequate for many construction purposes in rural areas, such as the construction of dams with earthwork cores (while higher qualities are used for constructions that demand more stringent standards).[38]

Since the Chinese approach (in contrast to the Indian) also allows for the use of small-scale labour-intensive cement technology in rural areas, it has an appeal that has not gone unnoticed. Spence, for example, has argued that '*a range of standards for cement should be instituted as in China, rather than the one single standard produced by the ISO*, and used almost everywhere else . . . A national system of quality control to ensure that all cement produced was correctly classified would also be needed. Further studies of how Chinese systems work in practice would be valuable background for such a policy.'[39]

The design of sanitation systems provides still another area in which the need for intermediate standards has come to be recognized. A comprehensive World Bank report on this topic, for example, warns against the search for a 'universal solution' to the waste-disposal problem in developing countries. The appropriate solution, according to this report, is much more likely to comprise

> . . . a combination of solutions adapted to different situations and communities. The fact that a universal solution is replaced by particular solutions should not be seen as a disadvantage. There is a danger that the main criticism leveled at any nonconventional waste-disposal system will be that it is not universal, that it will be difficult to implement, that new standards will have to be set, that personnel will have to be trained, and so on. All this is undoubtedly true, but it must be weighed against the disadvantage of a 'universal' solution that so far has been able to reach only 6.5% of the people in the developing countries.[40]

SUMMARY AND CONCLUSIONS

In both developed and developing countries the case for product standards rests on the failure of the unregulated market mechanism to provide sufficient consumer protection. However, the role of standards in developing countries may be expected to differ from that in developed countries in two major respects.

First, because of widespread illiteracy and the limited coverage of mass media, information is more difficult to convey to and be assimilated by consumers in the Third World. Consequently, provided that they lead to positive net benefits, product standards rather than warnings may be required to protect uninformed consumers in these countries. In the much publicized case of infant formula, for example, whereas the provision of information about the dangers of using these products may have only rather limited usefulness, the gains from a ban on their sale would be considerable. Not only could approximately 1 million lives a year be saved but also a vast amount of expenditure. The losers from a ban would comprise those, such as working mothers and women who cannot breastfeed, for whom infant formula is the optimal choice. But these categories are likely to be small relative to those who would benefit from the banning of formula. Thus according to Jelliffe: 'Studies in most slum or shantytown areas, in fact show that only a percentage (usually relatively small) are going away from the home to work.' Moreover, 'recent research shows that only a minute percentage of mothers are unable to feed their babies, even among the badly under-nourished.'[41] (To ensure that these categories do not suffer from the ban, infant formula could be made available under prescription.)

Secondly, though a case can be made for the imposition of more standards in developing than developed countries, there are also good reasons to believe that the standards adopted in the former should differ from those prevailing in the latter.

Contrary to a popular view we argued that, because the premium placed on additional safety by the consumer in rich and poor societies will generally differ, rich country standards tend to have a negative welfare impact when introduced into most developing countries. In the long run, as well, when embodied in international regulations or regulations inherited from the colonial period, such excessive standards may be inimical to the development of products which are suited to the needs of the majority of those living in the Third World.

In the case of bacterial standards for drinking water, for example, the WHO specifies international standards, which, while reasonable in developed countries, 'may well have been responsible for retarding the development of water supplies in less developed countries'.[42] With respect to drugs, legislation has in the past effectively exported US safety standards to the developing countries and, according to some commentators, this has hindered the development of new drugs which better conform to the needs of these countries.

DiRaddo and Wardell for example have argued for less-restrictive standards for the introduction of new drugs into human testing – that these be determined by the LDCs themselves – and for the 'expedited international transfer of molecules discovered in developed countries to LDCs for clinical testing'.[43] Further arguments in support of lower standards were cited in the areas of housing and building and of methods of sanitation.

In conclusion, we advocate that developing countries adopt intermediate standards, standards which require upgrading local norms on the one hand and downgrading developed country standards on the other. It is true that efforts to formulate intermediate standards will incur costs, including those of an administrative kind, which need to be offset against the arguments adduced above in support of a different set of standards. But the burden of the costs involved could be substantially reduced through cooperation among groups of developing countries with respect to, say, joint testing facilities, exchange of information and the formulation of standards.

Notes

1. World Health Organization (1971).
2. *Consumer Reports*, September 1970, p. 564.
3. For instance, Martin Khor of the Consumers' Association of Penang, Malaysia objected to the use of Depo-Provera (a contraceptive banned by the US FDA) on the grounds that it is 'grossly irresponsible' to 'distribute to our Malaysian women a drug which has been banned in the Western countries where it originated'. Quoted in the *Asian Wall St. Journal*, 10 April 1980.
4. Nath (1969), p. 9.
5. Festinger (1957).
6. The chapter entitled 'Why the Reckless Survive' in Konner (1990) contains a very readable summary of this literature.
7. Scitovsky (1976), p. 221.
8. Penz (1986), pp. 37–8.
9. Friedman and Friedman (1980), p. 227.

10. Colantoni, Davis and Swaminuthan (1976).
11. James (1977).
12. Ibid., p. 120.
13. Happold (1967).
14. Ibid. See also *The Economist*, 16 November 1991.
15. DiRaddo and Wardell (1979).
16. 'In a poor country like Mexico it is very bad when the breadwinner of a family gets sick. To get such a man back to work quickly, many physicians take a chance and prescribe a treatment that is a little dangerous. But if the patient is away from work too long, his wife and children may go hungry.' The example in the text is based on this quotation by a Mexican cardiologist quoted in Silverman (1976).
17. Jones-Lee (1976).
18. *The Economist*, 15 September 1990, p. 21.
19. Industrial Studies and Development Centre, Dar-es-Salaam, *Tender Document for a Turnkey Project for a Multi-Purpose Oil Mill Company*, 25 April 1975, p. 13, emphasis added.
20. Ibid.
21. *The Economist*, 15 September 1990, p. 22.
22. Ibid.
23. Temple and Temple (1980).
24. Ibid., pp. 235–6, emphasis added.
25. Ibid., p. 240.
26. Turner (1976), p. 75.
27. Hardoy and Satterthwaite (1990), p. 258.
28. Nelson (1979).
29. Ibid.
30. *The Economist*, 15 September 1990, p. 22.
31. Feachem (1980).
32. Kennedy (1979).
33. Eckstein (1961).
34. For a detailed discussion of water quality in developing countries, see Feachem, McGarry and Mara (1977).
35. Kennedy (1979), p. 191.
36. Silverman (1976).
37. *The Economist*, 15 September 1990, p. 24, emphasis added.
38. See Spence (1978) and Kaplinsky (1990).
39. Spence (1978), p. 52, emphasis in original. Hardoy and Satterthwaite (1990), p. 258, argue, more generally, the need for 'advice from centres, located throughout low-income areas, whose role would be to advise on how houses could be built or modified to improve health and safety. Government support for rapid expansion in the production of cheap building materials and building components, fixtures and fittings . . . could do much to lower the cost of building or upgrading a house . . . Support for workshops to produce such materials and fittings in each city district would also generate employment.'
40. Rybczynski, Polprasert and McGarry (1982), pp. 37–8. See also Sinnatamby (1990).
41. Jelliffe (1972), p. 201.

42. Feachem, McGarry and Mara (1977), p. 87.
43. DiRaddo and Wardell (1979), p. 276.

References

Colantoni, C., Davis, O. and Swaminuthan, M. (1976). 'Imperfect Consumers and Welfare Comparisons of Policies Concerning Information and Regulation', *Bell Journal of Economics*, Vol. 7, Autumn.

DiRaddo, J. and Wardell, W.M. (1979). 'Innovation and Availability in the United States of Drugs for Tropical Diseases', in US National Academy of Sciences, *Pharmaceuticals for Developing Countries*.

Eckstein, O. (1961). 'A Survey of the Theory of Public Expenditure Criteria', in Buchanan, J.M. (ed.), *Public Finances: Needs, Sources and Utilization*, Princeton University Press.

Feachem, R. (1980). 'Bacterial Standards for Drinking Water Quality in Developing Countries', *The Lancet*, August 2.

Feachem, R., McGarry, M. and Mara, D. (1977). *Water, Wastes and Health in Hot Climates*, Wiley.

Festinger, L. (1957). *A Theory of Cognitive Dissonance*, Stanford University Press.

Friedman, M. and Friedman, R. (1980). *Free to Choose*, Secker & Warburg.

Happold, F.H. (1967). *Medicine at Risk: The Higher Price of Cheap Drugs*, Queen Anne Press.

Hardoy, J. and Satterthwaite, D. (1990). 'New Partnerships for Healthy Cities', in Cairncross, S., Hardoy, J. and Satterthwaite, D. (eds), *The Poor Die Young: Housing and Health in Third World Cities*, Earthscan.

James, B.G. (1977). *The Future of the Multinational Pharmaceutical Industry to 1990*, Associated Business Programmes.

Jelliffe, D.B. (1972). "Commerciogenic Malnutrition', *Nutrition Reviews*, Sept.

Jones-Lee, M.W. (1976). *The Value of Life: An Economic Analysis*, Martin Robertson.

Kaplinsky, R. (1990). *The Economies of Small*, IT Publications.

Kennedy, D. (1979) 'Food and Drug Administration and Pharmaceuticals for Developing Countries', in US National Academy of Sciences, *Pharmaceuticals for Developing Countries*, Washington, DC.

Konner, M. (1990). *Why the Reckless Survive and Other Secrets of Human Nature*, Viking.

Nath, S.K. (1969). *A Reappraisal of Welfare Economics*, Augustus M. Kelley.

Nelson, N. (1979). 'How Women and Men Get By: The Sexual Division of Labour in the Informal Sector of a Nairobi Squatter Settlement', in Bromley, R. and Gerry, C. (eds), *Casual Work and Poverty in Third World Cities*, Wiley.

Penz, P. (1986). *Consumer Sovereignty and Human Interests*, Cambridge University Press.

Rybczynski, W., Polprasert, C. and McGarry, M. (1982). *Appropriate Tech-*

nology for Water Supply and Sanitation: Low-Cost Technology Options for Sanitation, A State-of-the-Art Review and Annotated Bibliography, The World Bank, Washington DC.

Scitovsky, T. (1976). *The Joyless Economy: An Inquiry into Human Satisfaction and Consumer Dissatisfaction*, Oxford University Press.

Silverman, M. (1976). *The Drugging of the Americas*, University of California Press.

Sinnatamby, G. (1990). 'Low Cost Sanitation', in Cairncross, S., Hardoy, J. and Satterthwaite, D. (eds), *The Poor Die Young: Housing and Health in Third World Cities*, Earthscan.

Spence, R. (1978). *Appropriate Technologies for Small-Scale Production of Cement and Cementitious Materials*, UNIDO Forum on Appropriate Industrial Technology, New Delhi, Nov. 20–25.

Temple, F. and Temple, N. (1980). 'The Politics of Public Housing in Nairobi', in Grindle, M. (ed.), *Politics and Policy Implementation in the Third World*, Princeton University Press.

Turner, J.F.C. (1976). *Housing By People: Towards Autonomy in Building Environments*, Marion Boyars.

World Health Organization (1971). *International Standards for Drinking Water*, 3rd edn.

6 Positional Goods, Conspicuous Consumption and the International Demonstration Effect Reconsidered

INTRODUCTION

Statements such as the following description of the West Indies abound in the development literature: 'With ready access to foreign goods and easy credit arrangements, an expanding bourgeoisie emulates the buying habits of the local elite and of more affluent societies . . . Exposure to metropolitan standards affects all levels of West Indian society . . . The external standards West Indians emulate include not only European but American modes of thought, behaviour, and material goods.'[1] Descriptions of this sort would seem to correspond to Nurkse's hypothesis of an international demonstration effect – namely, that 'the attraction of advanced consumption standards exerts itself fairly widely, though of course unevenly, among the poorer two-thirds of mankind'[2] – and at one level of course they do. But, what is seldom recognized is that Nurkse adopted a highly specific view of the mechanism which gives rise to this effect and as a result he was predisposed to a particular description of its consequences.

The purpose of this chapter is to argue, first, that Nurkse's view – especially his attempt to disassociate the demonstration effect from Veblen's notion of conspicuous consumption – is not at all well founded, because it is based on an excessively narrow vision of the process of taste and value formation in developing countries. On this basis, and second, it is contended that a reappraisal of the relationship between these concepts needs to be undertaken within a considerably broader socio-economic framework than that which Nurkse employed. Finally, the chapter is concerned to stress that the need for such a reappraisal is based on far more than mere theoretical curiosity; it derives instead from recent demonstrations by Hirsch, Scitovsky and Frank[3] that a Veblenian approach to consumer behaviour has welfare implications that are not only profound but also sharply different from those associated with the conventional theory of consumer demand. So far, none of these authors have considered the specifically developmental implications of their analyses – indeed all of them have confined their attention to developed market economies – but in so far as a Veblen-type approach has relevance also to consumption in the Third World, and to the demonstration effect in particular, equally if not more important policy implications may arise.

The first part of the chapter describes Nurkse's formulation and shows how the welfare implications of his view for individual consumers differ from those which would result from the alternative view of how individual preference functions are related. For this purpose, the product characteristics approach to demand theory is used. Following a discussion of the respects in which Nurkse's formulation of the issue is thought to be deficient, an alternative framework is proposed. Within the context of this alternative approach, the competing interpretations are then described and assessed.

NURKSE'S FORMULATION

The point of departure of Nurkse's formulation is that the preference functions of individual consumers in rich and poor countries are interdependent rather than independent. What is envisaged in particular is that 'When people come into contact with superior goods or superior patterns of consumption, with new articles or new ways of meeting old wants, they are apt to feel after a while a certain restlessness and dissatisfaction. Their knowledge is extended, their imagination stimulated; new desires are aroused.'[4] Nurkse is at pains to stress that this particular form of interdependence in consumption functions allows no room for any concept of social emulation through conspicuous consumption. As he himself put it:

> We can leave out Veblen's point that the propensity to spend is partly based on the desire for conspicuous consumption. I do not think that on the international plane the effect of unequal *living standards depends on the idea of 'keeping up with the Joneses.' All it depends on is demonstration leading to imitation*. Knowledge of or contact with new consumption patterns opens one's eyes to previously unrecognized possibilities. It widens the horizon of imagination and desires. It is not just a matter of social snobbishness.

> New products constantly emerge from the course of technical progress, which modify existing ways of life, and frequently become necessities.[5]

Before proceeding to describe the limitations of Nurkse's formulation and to show why it is not thought to be an adequate basis on which to judge the issue that is raised in the previous paragraph, it is necessary to demonstrate why the issue itself is important.

Why the issue matters

Use of the product-characteristics approach (which views products in terms of the characteristics that they embody) reveals that there are two distinct possible consequences for the welfare of the consumer in a poor country who is exposed to the demonstration effect *à la* Nurkse. These two possibilities – corresponding to what may be called a Galbraithian view and a Becker/Stigler household production function approach – are illustrated in Figure 6.1(a).

The figure assumes a world of two (indivisible) goods, two types of characteristics ('modern' and 'traditional') and prior to contact with the advanced countries, the individual in the poor country is consuming good *A* at point *P*. According to the one (Galbraithian) version of how Nurkse's effect operates, the knowledge that is transmitted (via advertising for example) alters the tastes of the consumer (in favour of modern characteristics), so that his entire indifference map shifts (from IC_1 to IC_2) and he moves to *Q* by replacing good *A* with good *B*. In this case, as is recognized in formal welfare economics, without a fixed frame of reference, it is impossible to assess the effect on the welfare of the individual consumer. (See, however, Figures 7.1 and 7.2 in Chapter 7.)

The alternative view of the individual welfare effect of contact with the developed countries originates in the household production function approach espoused by Stigler and Becker.[6] These authors argue that the effect of advertising (and presumably also other forms of the transmission of knowledge) is to alter the *information* available to the consumer, rather than his *tastes*. And since the welfare effect of the information that is conveyed depends only on its accuracy and in so far as the demonstration effect on balance improves rather than detracts from the knowledge available to the consumer, his choice of good *B* in Figure 6.1(a) will enable him to move on to the higher indifference curve IC_{11}.

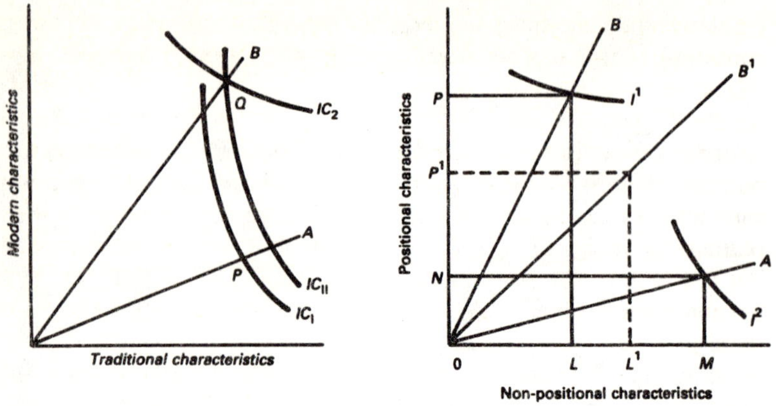

Figure 6.1 (a) Nurkse's view. (b) The Hirsch/Veblen view.

If the product characteristics approach is useful in revealing these two distinct variants of Nurkse's formulation, it is also helpful in elucidating the approach that he chose to eschew. In particular, it enables one to reconcile Veblen's early views on consumer behaviour with the similar ideas that have more recently been advanced by Hirsch. The latter argued that in so far as consumption is based on positional competition – that is, 'competition that is fundamentally for a higher place within some explicit or implicit hierarchy and that thereby yields gains for some only by dint of losses for others'[7] – it is collectively self-defeating. The fundamental problem is that '*The choice facing the individual in a market or market-type transaction in the positional sector, in a context of material growth, always appears more attractive than it turns out to be after others have exercised their choice.*'[8] Hirsch referred to goods whose consumption is based on positional competition as 'positional goods' or, more specifically, as 'those things whose value depends relatively strongly on how they compare with things owned by others.'[9] Conversely, non-positional goods 'depend relatively less strongly on such comparisons – the non-positional category includes, but is not limited to, goods that are not readily observed by outsiders.'[10]

Like Hirsch, Veblen believed that much of consumption behaviour is driven by 'emulation – the stimulus of an invidious comparison which prompts us to outdo those with whom we are in the habit of classing ourselves.'[11] Veblen also contended – and here a clear analogy with positionality begins to emerge – that the outcome of this emulative behaviour would be 'discrimination in favor of visible consumption',[12] a concentration of consumption 'upon the lines *which are most patent to the observers whose good opinion is sought.*'[13] And although he did not formulate the matter in the same rigorous zero-sum terms as Hirsch, Veblen nevertheless seemed to perceive that this conspicuous consumption was inherently economically wasteful, as the following quotation suggests. 'Relative or competitive advantage of one individual in comparison with another does not satisfy the economic conscience.'[14] Unlike Hirsch, however, Veblen maintained that it is the characteristics of goods and not the goods themselves that are required for 'positional' or 'conspicuous' purposes and he suggested, furthermore, that these are rarely the *only* reasons why particular products are in demand. In general, that is to say, the demand for goods has *both* a positional and a non-positional component. Thus:

> It is obviously not necessary that a given object of expenditure should be exclusively wasteful in order to come in under the category of conspicuous waste. An article may be useful and wasteful both, and its utility to the consumer may be made up of use and waste in the most varying proportions. Consumable goods, and even productive goods, generally show the two elements in combination, as constituents of their utility.[15]

By combining the ideas of Hirsch and Veblen it is possible to represent the welfare implications of this synthesis – the type of approach that Nurkse chose to reject – in terms of Figure 6.1(b). Unlike Figure 6.1(a), the two goods *A* and *B* are now defined by their embodiment of positional and non-positional, rather than modern and traditional, characteristics. The individual initially chooses good *A* but given his new tastes, which, because of the demonstration effect, reflect the increased orientation to the positional aspect of demand, he changes to good *B*. Following Hirsch, however, the positional characteristics turn out (*ex post*) to be entirely illusory. That is, 'after others have exercised their choice', all the consumer ends up with is $0L$ of intrinsic or non-positional characteristics, which is less than was obtained ($0M$) with the original preference ordering. It is worth noting that the extent of the absolute decline in non-positional characteristics (and hence the degree of the consumer's *ex post* loss in welfare) depends on the nature of good *B* (i.e. on the slope of the ray representing this good). If, instead of *B*, the consumer had chosen good B^1, for example, it is clear that the diminution of non-positional characteristics would have been much less severe. We shall explore the welfare implications of this point below, when a detailed interpretation of the demonstration effect along Hirsch/Veblen lines is attempted. For the present purpose, it is sufficient to note that if this effect is represented as causing an enhancement of positional tastes, the welfare impact is quite different from that which is associated with Nurkse's view.[16]

The limitations of Nurkse's formulation

If the issue raised in this chapter seems therefore to bring up a potentially important question of welfare, it is also fairly plain that Nurkse's formulation – and in particular his dismissal of a Veblenian interpretation – is inadequate. For it entirely ignores the *manner* in which individuals *become* responsive to, and predisposed to search for, the knowledge about advanced country products that is transmitted through the demonstration effect. And it also fails to address the question of how, in the process by which tastes for *products* are altered, *other* proclivities of individuals may also be affected. Yet, these factors are likely to constitute essential determinants of the type of interdependence between individual utility functions that is created by the process that Nurkse describes (i.e. whether the process is better described by Figure 6.1(a) or (b). A more adequate assessment needs to incorporate these omitted factors in a much wider analysis of how tastes and values are formed in developing countries. And this in turn means that one has to enlarge the scope of the enquiry beyond the confines of economic variables (with which Nurkse was mostly concerned) to include the findings of sociologists, and particularly those sociologists who have studied the modernization process.

MODERNIZATION THEORY

As it is formulated by modernization theorists, the sociological approach differs from that taken by economists not merely with respect to the specification of the independent or explanatory variables but also (as Table 6.1 indicates) with regard to the range of variables that are to be explained.

Whereas in the former approach changes in taste are *isolated* from

Table 6.1 Economic vs. sociological approaches to taste change

	Economics	*Sociology (modernization theory)*
Independent variables	Income (absolute and relative), advertising	Urbanization, factory employment, education, mass-media exposure, etc.
Dependent variables	Taste change in isolation	Value change, openness to new ideas, ambition, secularism, participation, etc. (the 'modernity syndrome')

any other accompanying (psychological and socio-cultural) changes, sociologists tend to a view in which all these various dimensions of change occur *simultaneously*, and often inextricably, as part of what is referred to as a 'modernity syndrome'. In particular, together with the acquisition of a taste for modern goods, 'modern man':

> . . . identifies with the newer, larger entities of region and state, takes an interest in public affairs, national and international as well as local, joins organizations, keeps himself informed about major events in the news, and votes or otherwise takes some part in the political process. The modern man's sense of efficacy is reflected in his belief that, either alone or in concert with others, he may take actions which can affect the course of his life and that of his community; in his active efforts to improve his own condition and that of his family; and in his rejection of passivity, resignation, and fatalism toward the course of life's events. His independence of traditional sources of authority is manifested in public issues by his following the advice of public officials or trade union leaders rather than priests and village elders, and in personal matters by his choosing the job and the bride he prefers even if his parents prefer some other position or some other person. The modern man's openness to new experience is reflected in his interest in technical innovation, his support of scientific exploration of hitherto sacred or taboo subjects, his readiness to meet strangers, and his willingness to allow women to take advantage of opportunities outside the confines of the household . . . *In other words, psychological modernity emerges as a quite complex, multifaceted, and multidimensional syndrome.*[17]

What is striking about the vector of factors that make up the syndrome described in the quotation is not only that it seems to appear consistently in studies that are independently conducted, but also in a variety of diverse populations. And because some sociologists have devoted a considerable amount of time to the empirical investigation of the factors that contribute to the 'modernity syndrome', much can be learnt about the causes of value change (which, in the transition from traditional to modern values, comprise an integral part of this syndrome) from an analysis of their results. This knowledge in turn, ought to throw a good deal of light on the assessment of the competing interpretations that were described above.

The determinants of modernity

Using a summary measure of modernity that incorporates a very wide range of attitudinal variables, Inkeles and his collaborators sought to identify the factors that were most important to explaining the variation in this measure among a large sample of individuals in six developing countries. What they found was that approximately 50 per cent of the variance in the data could be accounted for by a set of eight variables of which education, experience in a modern large-scale factory and exposure to mass media were the most important. But modernity also appeared to be influenced to an important degree by the *country* in which these separate influences were exerted. In particular, the researchers:

> . . . found no way to escape the fact that the men from some countries consistently scored higher in individual modernity than those from other countries, even when the individuals compared were apparently alike in certain characteristics which had previously been shown to be the most powerful determinants of such scores . . . A man from Argentina who had completed more than eight years of schooling and a Chilean with about seven years of schooling did only as well as an Israeli who had only been to school for three years.[18]

The authors concede that the reasons for this 'national character' effect on modernity are not at all clear but suggest that the most likely explanation has to do with a view of modernization in which 'individuals living in a more modern setting should become more modern merely by sharing a *generally* modern ambience. And one important factor in making a modern ambience may be the *average* level of modernity of the individuals who live in the environment.'[19] Because this average is also highly correlated with more objective indices of development such as GDP per head, 'the richer countries will enjoy a double advantage.' For not only will they be better able to supply more of the factors (such as education, factory jobs, etc.) that contribute to modernity but their very affluence *in itself* contributes to the process.

The social learning view of taste change

Earlier we noted that one reason for the inadequacy of Nurkse's formulation is its failure to provide insight into the question of why individuals become favourably disposed towards or inclined to search for the knowledge that is conveyed about developed-country products via the demonstration effect. The evidence presented in the previous section indicates the variables that are important to this transformation, but provides no view of what sociological processes underlie the explanatory variables.

Inkeles and his collaborators assert that the causal factors underlying their results (as well as those of other modernization theorists) mostly derive from the principle of 'social learning'. That is, in their view individuals become modern by internalizing the values which are 'embedded in the organizational practices of the institutions in which they live and work.'[20]

Consider, for example, the way in which education and work in large-scale factories are thought to promote modern attitudes according to this approach (which derives its rationale principally from the Marxist view that it is the relationship to the mode of production that is fundamental in shaping one's consciousness). Thus:

> School and factory produce the same result because they both expose individuals to certain common principles of organization, procedures for assigning power and prestige, modes for allocating rewards and punishment, and approaches to the management of time. Individual modernity then becomes a quality learned by the incorporation into the self-system of certain qualities characteristic of particular institutional environments.[21]

What is stressed by this approach, therefore, is the general, rather than the specific nature of the impact that is imparted by the modernizing institutions. In the case ot education, for example, the process by which the school inculcates modern values has mostly to do with 'the distinctive nature of the social organization as such' and little to do with 'the curriculum as such'.[22] As an example of how the *inherent* characteristics of this institution work to produce this effect, consider its relation to the notion of 'a sense of efficacy', which, as noted earlier, is one component of the 'modernity syndrome'. Thus:

> In the acquisition of this feeling *generalization* plays a substantial role. Generalization occurs when an individual enjoys so satisfying an experience in one specific relationship or performance that he is led to believe that he can attain comparable success in other contexts . . . The child who learns to read his school books later finds himself able to read directions and instructions and to follow events in the newspaper . . . By extension and diffusion, or what we have called generalization, a heightened general sense of personal efficacy results.[23]

Or, to take another example, it is argued that 'principles directly embedded in the daily routine of the school teach the virtue of planning ahead and the importance of maintaining a regular schedule.'[24]

It follows from this social learning view that there is an essential similarity in the way that individuals become modern, *regardless of the context in which the process occurs*. Indeed, even the vast diversity that is brought about by different economic systems (capitalist vs. socialist), cultural and political variations and so on, are thought not to have a 'great impact on the process of individual modernization. The critical factor in that process will be the extent to which individuals are exposed to the "schools for modernity" such as formal education, factories, modern offices.'[25]

In brief, then, according to the social learning view the 'emergence of modernity is primarily a consequence of *internal dynamics* of each society rather than of external forces. With law-like regularity, the same psychological traits arise in different and relatively independent national settings in response to the *same* structural pressures. Modernity is the natural response of personality and culture to incipient processes of structural transformation wherever they may occur.'[26] In so far as it is valid, this view lends support to Nurkse's formulation of the demonstration effect for it shows how individuals can become predisposed to favour advanced country products *in a way that need have no Veblenian positional component at all.*[27] However, as Portes has pointed out, even on the basis of the same explanatory variables an alternative interpretation of the modernization process is possible. In particular:

> as an alternative to the dramatic portrait of a new personality system arising spontaneously and independently in different social contexts – the universality of modernity is an obvious consequence of the universality of diffusion of modern, i.e. Western, values and behavior patterns. Many people in many countries have had the opportunity to be socialized in, essentially the same set of orientations . . . Massive aspirations, media-consumption and other such modern traits do not arise naturally from *internal* processes of structural change, but *artificially* from the impact of Western cultural diffusion centered in urban areas. *Factors which have been repeatedly found to promote modernity – education, industrial occupation and urban residence – may function less as carriers of self-evident psychological implications discovered anew by each developing population than as vehicles for closer and more intensive exposure to Western values and lifestyles.*[28]

It is on the basis of this view that one can formulate an alternative interpretation of the implications of the demonstration effect – an interpretation that will rely upon, rather than eschew, the ideas of Veblen and Hirsch.[29]

AN ALTERNATIVE INTERPRETATION OF THE DEMONSTRATION EFFECT: THE TRANSFER OF POSITIONAL VALUES

Following Maslow's[30] concept of a hierarchy of needs, the striving for status (or the acquisition of high standing relative to others) becomes a predominant motivation only after other, more basic needs (for hunger, thirst, safety and belongingness) are satisfied.[31] In this sense, Hirsch's positionality or Veblen's 'conspicuousness' may be described as a 'high-income' taste that is indulged principally by those with incomes sufficient to have satisfied their more fundamental needs. Yet, according to the alternative interpretation of the demonstration effect that this section will present (that corresponding to Figure 6.1(b)), it is precisely this high-income taste that (among other things) is transferred by the demonstration effect to the poor countries of the Third World. This alternative formulation differs from that described in the previous section not in its view of the set of variables that are thought to be important to the modernization process, but rather in the manner in which they are said to contribute to this process. It is to the elucidation of this crucial distinction that we now turn.

The instruments and mechanism of the transfer of positional values

Education and work in large foreign factories

In the alternative view the role of education and work in large foreign firms is to transfer developed-country values, including positional values, to developing countries. Wilson's social anthropological study of the English-speaking societies of the Caribbean provides a fascinating example of this view. Wilson found that for these communities, 'The reference group – for high social status and for respectability – is the urban middle-class Afro-American and white community of the United States or the Panama Canal Zone. This is the model where proper English is spoken, where manners come naturally, where a house is immaculate and furnished to the best taste to the utmost convenience, where sophistication has its roots and modernity its zenith.'[32] Wilson shows how this role orientation produces a pattern of consumption that can only be described as highly positional. For instance, 'Economic differentiation becomes a public and social matter through the assumption of a life style, marked in part by the acquisition, use, and display of goods . . . Economic differentiation is clearly a large and vital part of social differentiation – the reclassification of people in terms of their status.'[33] Or again, and more specifically, 'Respectability at its material level, at its level of signification, is a show of luxury. It is a grand, well-furnished home, well equipped with modern appliances, fine furniture, china and linens, good stylish clothes, an expensive education, manners and deportment. These signifiers are also the foci of ambition for the population as a whole.'[34]

What drives the patterns described in the previous paragraph according to Wilson is primarily the system of education. 'Euro-American ideas and ideals enter the society most pervasively and influentially through the educational system. Here they are instilled, often obliquely, as the standards of right and wrong. It is here that youthful hopes and ambitions are engendered and directed – but are all too often left unfulfilled.'[35] A similar role appears to be played by the church and:

> in the professional standards and moral expectations that Euro-American institutions, firms, hotels, and employers impose on their employees, beginning with standards of dress and address and going deeper by nurturing ambitions and setting the terms by which those ambitions might be realized, including, a 'respectable' outlook on life and business. To the extent that companies and employers do this, *they invade the value system of the Caribbean*, and the more powerful they are, the more successful their invasion.[36]

More generally, Sauvant has suggested that:

> the ability of foreign affiliates to influence and shape a host country's business culture – and through it the entire cultural and social system – is based on two factors: first, on their direct capacity to add to and shape the production apparatus of the host country and to introduce, promote, and disseminate new modes of operation and behavior; and, second, on the impetus they give to the creation of a supporting (foreign-controlled) business service structure geared to meet the needs of foreign affiliates. The crucial common element in these factors is that the capital investments underlying international business are accompanied by sociocultural investments, that is, values and behavioral patterns associated with the parent enterprise (or agency) and its home country.[37]

Advertising

Advertising is likely to contribute to an enhanced preference for the positional component of demand for two reasons. First, as Hirsch points out, it tends to 'encourage the strengthening of self-regarding individual objectives and makes socially oriented objectives more difficult to apply. The reason is that interests of self-concern and self-regard can be enlisted much more effectively in support of commercial sales efforts.'[38] Second, the appeal to self-interest is often made with direct reference to positionality – that is, through linking consumption of the advertised product to 'getting ahead' of the rest of society. Together, these features of advertising foster not merely the existence of an individualistic ethic (which would remain true, incidentally, even if only information is conveyed), but also the notion that self-interest is best pursued with respect to advancing one's rank order in society.

In developing countries much of the advertising and sales promotion that is undertaken by multinational corporations remains essentially unaltered from its original, host country form. As a result, these techniques are likely to have the same effect on the positional component of values that was described in the previous paragraph. For example, in the context of the promotion of tonic foods in the Third World, Jelliffe has noted that 'The content of advertising and the techniques used are those widely employed in the Western World, with emphasis on status and convenience. The tropical mother is in many cases even more vulnerable to these motivating forces . . . She is as anxious to emulate the well-to-do elite, both local and foreign, as is any Western mother.'[39] A very similar process is described by Ledogar[40] in the case of the marketing of soft drinks.[41]

Direct historical contact: the impact of colonialism

A study of occupational prestige and social status in (what was then) Northern Rhodesia by Mitchell and Epstein provides some interesting observations on the impact of colonialism on urban consumption patterns and in particular on those that might be described as positional. The authors point, for example, to the acceptance in urban communities of the prestige scale introduced by the Europeans. 'In other words, the social grading of occupations reflects the more generalized prestige system which manifests itself as the emulation of the way of life of the socially dominant Europeans . . . *Success in achieving this "civilized" way of life is demonstrated conspicuously by the physical appurtenances of living*. The most important of these is clothes, but personal jewellery (especially wrist-watches), furniture, and European-type foodstuffs are also important.'[42]

It is important to note that the propensity for this kind of positional consumption in the urban (as opposed to rural) areas of developing countries may derive not merely from more numerous contacts with the developed world (the factor that in Nurkse's formulation endows the city with its modernizing influence), but also from pressures that are inherent in the process of urbanization itself. In this regard Veblen himself believed that 'the serviceability of consumption as a means of repute, as well as the insistence on it as an element of decency, is at its best in those portions of the community where the human contact of the individual is widest and the mobility of the population is greatest.'[43] For this reason he asserted that 'Conspicuous consumption claims a relatively larger portion of the income of the urban than of the rural population and the claim is also more imperative.'[44] It is also worth noting his insistence that 'It is not that the city population is by nature much more eager for the peculiar complacency that comes of a conspicuous consumption, nor has the rural population less regard for pecuniary decency. But the provocation to this line of evidence, as well as its transient effectiveness, are more decided in the city. This method is therefore more readily resorted to, *and in the struggle to outdo one another the city population push their normal standard of conspicuous consumption to a higher point.*'[45]

Urban patterns of conspicuous consumption may also differ from rural with respect to the *type* of goods that are chosen for this purpose. These differences can be nicely illustrated with examples drawn from Epstein's study of two villages in South India. Both had undergone substantial economic development in the 25 years prior to Epstein's study in 1962.[46] One village (called Wangala), however, had remained wholly agricultural while the other, Dalena, had become more closely integrated with the regional and commercial system. These differences, according to Epstein, were reflected in very different forms of status-seeking behaviour. In particular:

> Dalena's economic diversification led to greater economic differentiation and thus to the development of more and more refined criteria of prestige, arising out of the contact with the town. Dress has become an important criterion of prestige . . . The richest Dalena men have to dress better in order to impress officials in the administration in Mandya [the town] . . . The two Dalena men who are employed as clerks in Mandya wear western-style suits when they go to work. Thus the prestige accorded by dress arises out of the different economic activities performed by Dalena men. In Wangala rich and poor alike work on the land and use tattered cotton shirts and shorts for everyday wear. Here uniform occupation has prevented dress from becoming an important criterion of prestige except on special occasions.[47]

Or, to take another example:

> Increased expenditure on weddings in Wangala expresses a concern for prestige in a closely-knit community. By contrast Dalena villagers have so many different links with surrounding villages and

Mandya that the struggle for status within the traditional, closely integrated, village system has lost some of its importance to them. In their eyes, items which are readily transportable, and which can easily be shown off outside their own village, award prestige. These items, such as dress, pens, pencils, gold watches, or bicycles, will yield prestige to their owner in a wider sphere of social contacts and not just in his own village. Accordingly, rather than introduce novel items into the traditional wedding procedure, which after all would yield prestige only among a very limited number of people, Dalena villagers prefer to spend money on purchases of items which they can display easily on visits to neighbouring villages and to Mandya.[48]

The welfare consequences of positional taste transfer[49]

So far in this section fragments of evidence have been adduced in support of the interpretation that the demonstration effect occasions a change towards positional values and behaviour on the part of (some of) those who live in the Third World. But as Figure 6.1 (b) was concerned to demonstrate, the welfare effects of this change depend on the type of good that is used *to give effect to* the new values. And in order to get some sort of analytical handle on this issue it is clear that one needs to convert the concepts of positional and non-positional characteristics (representing the two axes of Figure 6.1 (b)) into more concrete categories and to investigate how these categories are combined into the different goods that are available for the satisfaction of enhanced positional values.

Mapping positional demand into its constituent categories of product characteristics

We have already taken note that visibility is a part of the *definition* of positional consumption and it is also a characteristic that forms an essential part of the promotional activities (packaging, advertising and brand differentiation) of multinational firms in developing countries. But though it is necessary, visibility is not a sufficient characterization of positional demands. It requires, as an essential complement, the quality of 'superfluousness'. In Veblen's words, 'Throughout the entire evolution of conspicuous expenditure . . . runs the obvious implication that in order to effectually mend the consumer's good

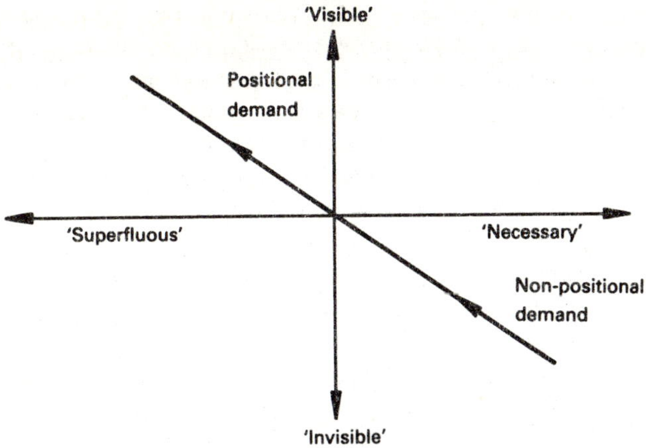

Figure 6.2 Positionality and the direction of change in demand.

fame it must be an expenditure of superfluities. In order to be reputable it must be wasteful. No merit would accrue from the consumption of the bare necessities of life.'[50] Or again, 'The consumption of expensive goods is meritorious, and the goods which contain an appreciable element of cost in excess of what goes to give them serviceability for their ostensible mechanical purpose are honorific.'[51]

Taken together, these two components of positionality imply the direction of change in the demand for characteristics that is broadly indicated by the north-westerly movement from the second to the fourth quadrant of Figure 6.2.

The question that now needs to be addressed in assessing the welfare implications of the transfer of positional values is the nature of the means that are available to consumers to obtain the bundle of characteristics represented in the fourth quadrant of the figure. More specifically, one is required to ask whether and to what extent this bundle can be obtained without a sacrifice in other, non-positional characteristics (for it is this that determines whether the consumer reaches L or L^1 in the example of Figure 6.1 (b)).

Two general considerations would appear to be of special significance in answering this question. The first has to do with the historically distinctive condition of contemporary underdevelopment and in particular the near total dependence of developing countries on goods developed in and for advanced countries. New goods in the latter, whose members on average enjoy a rising standard of living, come to acquire an increasingly high proportion of 'high-income' characteristics over time (including those such as sophisticated packaging, advertising and brand differentiation that may be described as highly visible) and a lower balance of low-income characteristics.[52] In so far as these developments continuously cause existing products to become outmoded[53] it follows that consumers in poor countries are able to buy the bundle of characteristics that we described as positional, *at an increasingly high cost in terms of non-positional characteristics*.

The second general consideration has to do with the closeness of the link that often exists between the way in which positional values are created and the manner in which they need to be met. The promotional activities of multinationals, for example, may *simultaneously* alter values in favour of positionality and require the new values to be met by the specific products that are the subject of promotion. To the seemingly frequent extent that these products are transferred unaltered from the form in which they are sold in developed countries (e.g. Coca-Cola, infant formula), the result is again that in order to acquire the package of positional characteristics, a high price has to be paid in terms of foregone non-positional characteristics.

These two general arguments amount to positing a substantial *discontinuity* in the process by which the consumer is able to give effect to the change in demand depicted by (the direction of the arrows in) Figure 6.2. This discontinuity, and the welfare loss with which it is associated, bears contrast with the more continuous process that seemed to occur during the historical experience of the now developed countries. Towards the end of the nineteenth century in the United Kingdom, for example, when powerful pressures to emulation were given by increasing urbanization, the growth of advertising and mass media and the awareness of social class, 'There could have been no large shift towards the range of "new" commodities associated with the "high mass consumption" of the twentieth century.'[54] Emulative desires appear instead to have been indulged at that time through a relatively continuous or smooth process of enhancement to traditional patterns of consumption: 'new instead of second-hand clothing, leather footwear, gas instead of candles, coal instead of wood, commercial instead of home-made soap, improved types of linoleum floor-covering, arm chairs instead of kitchen chairs'.[55]

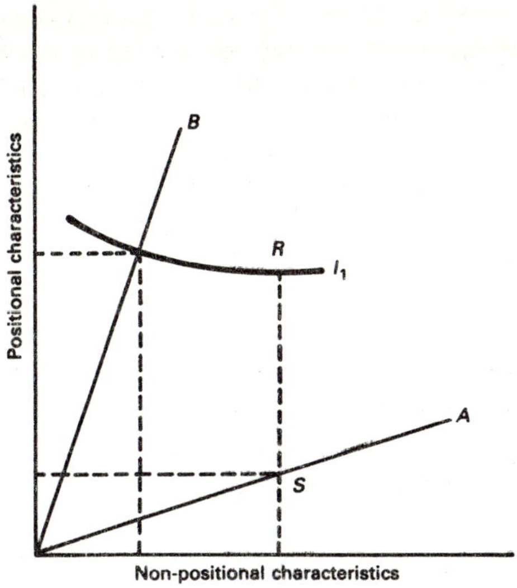

Figure 6.3 The appearance of modernity.

Giving the appearance of modernity

So far, our argument has been concerned with the type of embodiment of the characteristics combination that was earlier defined as positional. We have tried to show that the range of goods available is rooted in the historical conditions of 'latecomer development' and that the welfare implications of positional consumption in the Third World have to be understood in these terms. Implicit in our analysis, however, is the assumption that the positional package of characteristics can only be obtained by consumption of a more 'positional-intensive' good. To some (unknown) degree, this assumption will not hold because consumers are able in some circumstances to acquire the positional characteristics of goods without actually consuming the goods themselves. That is, they merely give the *appearance* of so doing. Consider, for example, the situation described by Figure 6.3.

Goods *A* and *B* represent two alternative writing instruments: respectively a pencil and a ballpoint pen. If the consumer were to switch to the latter because of its higher balance of positional elements, the welfare consequences would be those described earlier in relation to Figure 6.1 (b). Let us assume, instead, however, that the consumer is able to acquire a discarded ballpoint pen top, which, when inserted into his pocket, effectively gives the appearance of consumption of this good. If we assume, further, that this ploy enables him to obtain an amount equal to *RS* positional characteristics, he can reach the indifference curve I_1 while continuing to consume good *A*. In this case, therefore, the change towards positionality leads to no loss in welfare. (That is, for the consumer himself. In so far as they purchase *complete* ballpoint pens for positional reasons, other consumers will tend to be adversely affected.)

That something like the process described in the previous paragraph actually occurs in developing countries is very nicely illustrated by the following description of the variation in the choice of building materials in a region near Buenos Aires:

> The principal variation was constituted by the nature of building material, the poorest huts being made of mud mixed with straw and reeds, the 'richest' being bricks and mortar. In between these two extremes . . . there was a series of combinations in which *efforts were made to give the impression of brick construction, either by white-washing the mud walls in a way that made them appear to be rendered, by laying the bricks in mud rather than mortar and subsequently white-washing them, or by constructing the front wall which faced the roadway of bricks, and making the others, which were out of public view of mud.*[56]

Other examples from Africa, include 'the broken wireless or gramophone, the sunglasses without lenses and the shirt with no back'.[57]

Interestingly, cases such as these closely resemble the phenomenon that Grossman and Shapiro[58] refer to as the foreign counterfeiting of status goods. What they mean by this are purchases made knowingly of counterfeit versions of luxury products such as (Gucci) handbags and (Chanel) perfume, at prices considerably lower than the authentic versions of these products. Grossman and Shapiro cite (what seems to be fairly reliable) evidence to suggest that the phenomenon is far from unimportant. In fact, the shares of fake designer sunglasses and watches in the United States market may be as high as 25 per cent, while the share of counterfeit jeans could amount to something of the order of 10 per cent.

Of the issues raised by these estimates, the most important for our purposes is the light that they throw on the extent of positional consumption in the developed countries. For, given the typically low quality of the counterfeit products, one has to assume that they are acquired primarily for the prestige that they (falsely) bestow on those who consume them. In fact, when they are acquired entirely for this purpose, they become identical to the examples cited earlier (in the developing country context) that confer status by giving the appearance of modernity. And as in the latter, counterfeits reduce the welfare cost of engaging in positional consumption because they make available the positional characteristics of products, independently of other functional characteristics. In both cases, however, a welfare loss is imposed on those who consume the 'genuine' versions of the products because the exclusiveness of these is diminished by the presence of the illegitimate versions.

Low-income positional consumption and the squeeze on essential characteristics

In so far as the device of giving the appearance of modernity is limited (and consumers have therefore to purchase good *B* in Figure 6.3 in order to satisfy their positional demands) and to the degree that the lowest deciles of the population in developing countries indulge in positional consumption, the problem of a 'squeeze' on low-income, or essential, characteristics is raised. In a different context, and in somewhat different form, this problem was of concern also to Veblen because of his belief that 'No class of society, not even the most abjectly poor, foregoes all customary conspicuous consumption. The last items of this category of consumption are not given up except under stress of the direst necessity.'[59] As a result, and especially in certain lines of consumption, he perceived that the outcome would be pressure on the essential elements of consumer demand. In particular, 'It is true of dress in even a higher degree than of most other items of consumption, that *people will undergo a very considerable degree of privation in the comforts or the necessaries of life in order to afford what is considered a decent amount of wasteful consumption; so that it is by no means an uncommon occurrence, in an inclement climate, for people to go ill clad in order to appear well dressed.*'[60] Frank too believes that positional consumption will tend to be most intense among the poor. The reasons he gives for this are two:

> First, just as the poorly fed person derives more satisfaction than others from additional food, so too will the low-ranked person derive greater satisfaction from an advance in the rankings.
>
> The second reason is that any given positional expenditure will

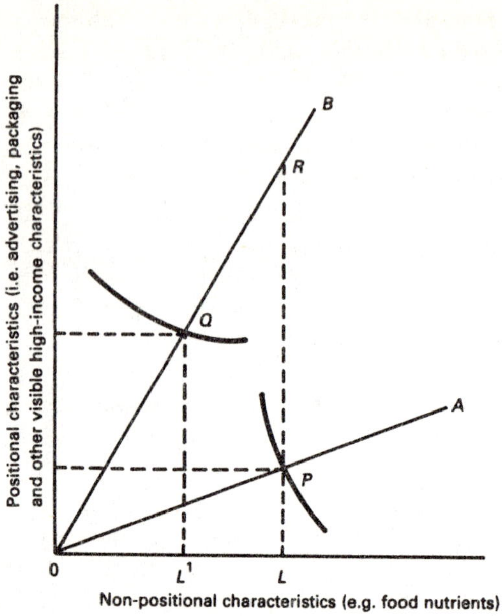

Figure 6.4 The squeeze on essential characteristics.

generally produce a larger change in rank for low-income than for high-income persons. This happens because people are bunched together more closely nearer the bottom of the economic ladder. For a low-income person, there are many people with consumption levels only slightly higher than his own, so even a small increase in positional expenditure will appreciably increase his rank. But in the upper reaches of the income distribution, where incomes are more spread out, the same expenditure will not affect rankings by much. A hundred dollars more spent on clothing, a hundred dollars less saved – all such expenditure shifts will produce larger advances in the ranking for persons near the bottom of the hierarchy than those near the top.[61]

For the poor consumer in a contemporary developing country, the problem that gives rise to the squeeze on essential characteristics can be described as in Figure 6.4.

Following the alteration in his tastes between positional and non-positional characteristics, the low-income consumer faced only with modern good *B* imported from the advanced countries suffers a decline in essential characteristics equal to L^1L as a result of the consumption of this good. Only with an income level that would enable him to reach point R along the ray representing the modern good, would he be able to maintain the original level of essential characteristics $0L$.

The extent to which the process described in Figure 6.4 actually occurs is impossible to assess on the basis of available data. The most that can be pointed to is fragmentary evidence that the process seems to occur primarily in urban areas of the Third World. In the case of infant formula, it seems to be due in part to promotional activities that extend even to the poorest groups[62] and in part to emulation by these groups of the positional values and behaviour of the relatively affluent.[63] In the case of food advertising, a process very similar to that described in Figure 6.4 has been observed by nutrition experts. The following quotation is perhaps the clearest statement of this view of the consequences of advertising of imported food products:

> even with their limited means, the poor buy highly preferred or 'fashionable' foods at a premium in price. The urban poor may buy them for their status value, although they may be uneconomical from the nutritional standpoint. Foods that have low prestige in rural areas, such as fruits and vegetables which are gathered and not cultivated, may not be eaten by newcomers to urban areas because they cost money and are not thought of as 'valuable.' On the other hand, foods may be eaten for status reasons.
>
> An example of such behaviour is the purchase of expensive commercial formula products, and other 'prestige' foods such as bread, soft drinks, sugar, tea, infant formulas, and canned milk. The acquisition of these foods by the poor may have a nutritionally detrimental effect due to the economic drain they place on the meager food budgets.[64]

The interstatus income ratio compression effect of positional consumption by the poor

In the degree to which positional consumption by the poorest groups in developing countries does take place in the manner described above, the result, paradoxically, may be to induce additional consumption of this kind by the *richest* groups in these societies. This effect, which has been well-documented by labour economists, and

Table 6.2 The interstatus income ratio compression effect and positional consumption

	Period 1			*Period 2*		
Status	*Income*	*Positional consumption*	*Non-positional consumption*	*Income*	*Positional consumption*	*Non-positional consumption*
	(1)	(2)	(3)	(4)	(5)	(6)
(Lowest) 1	100	50	50	200	100	100
2	200	100	100	300	200	100
3	300	150	150	350	300	50

which has been used by Leibenstein to explain fertility decisions, can most easily be explained with the aid of Table 6.2 (adapted from Leibenstein, 1975, p. 5).

Table 6.2 shows a status hierarchy – the mean income of each successive status group is higher than its predecessor. Three status groups are assumed in a two-period framework (the two periods may be viewed as being roughly 15 years apart). The interstatus income ratio compression effect is shown by the numbers in column 4 compared to those in column 1. 'For example, in a developing country an engineer might earn 20 times the income of an agricultural laborer, while in a high-income country the ratio might be reduced to three.'[65]

It is further assumed that the income distribution remains roughly constant as per capita incomes rise. 'Thus, if households move disproportionately to higher statuses between the two periods, and simultaneously interstatus income ratios fall, then it is possible to retain approximately the same income distribution. The relative income compression effect will be compressed all the more if the income distribution becomes more equal as development occurs.'[66]

By making it easier for low-status households to emulate higher ones, the compression effect forces the latter to increase sharply positional expenditures in order to maintain status differentials. This squeeze on non-positional expenditures in period 2 is illustrated in Table 6.2. If those in status 1 spend the same amount on positional consumption in the second period as did status 2 families in the first period, then status 2 families have to double their positional expenditures to maintain the differential (i.e. a ratio of 2 to 1).[67]

CONCLUSIONS

That spending patterns in poor countries are susceptible to influence from the consumption behaviour of the richer societies has always seemed self-evident. But just how this influence is transmitted and with what effect on the welfare of the individual consumer in the Third World are questions that admit of much less certainty. In a well-known contribution in the late 1950s, Ragnar Nurkse suggested that the process works through an 'international demonstration effect'. By this he meant essentially the exposure of the poor to modern products – 'demonstration leading to imitation' – and he explicitly eschewed the notion that international emulation of the rich had anything to do with status seeking, or, to use Veblen's phrase, with conspicuous consumption.

This chapter has argued that there are a number of reasons why Nurkse's contribution needs to be reassessed. First, in the light of recent analyses of the advanced countries (by Hirsch, Frank and Scitovsky) showing the powerful welfare implications of consumption based on status seeking, the issue of whether and to what extent this form of consumption behaviour also applies *internationally* takes on a much greater significance than might earlier have seemed to be the case. Second, it is contended that Nurkse's approach is inadequate to appraise this important question and that a much broader framework – encompassing sociological as well as economic variables – is required.

On the basis of this expanded framework, it was shown that underlying the different views of the way in which preference functions are related at the international level is a set of quite distinct hypotheses concerning the influence of a range of socio-economic variables on the formation of values and preferences in developing countries. Although a fairly substantial body of sociological and other evidence was brought to bear on the assessment of these competing causal mechanisms, and though there seemed sufficient material to cast doubt on Nurkse's eschewal of status-seeking consumption behaviour at the international level, a clear verdict was not able to be reached. In part, this lack of conclusiveness is due to the inherent difficulty of conducting (and for this reason also the lack of) research that is able to elicit the 'true motivations' of consumers. It is now known from social psychology, for example, that direct and indirect methods of uncovering human motivation sometimes produce different results.[68]

But the lack of conclusiveness is probably also a reflection of the fact that the competing views should be regarded less as universal truths than as descriptions of consumption behaviour that are applicable to different socio-economic circumstances. Much may depend, for example, on the extent to which group values are retained to exert a countervailing influence on positional consumption behaviour during the modernization process. The importance of the group in this respect has been emphasized by Douglas and Isherwood:[69]

> A strong group has its own characteristic ways of controlling the envy that might spoil the relations of its members and so threaten its permanence.
>
> The group imposes group values and so prevents deviant individual spending, defines what counts as too much conspicuous individual consumption, and proposes punishments.[70]

The Japanese experience represents probably the clearest example of the workings of a social (the 'Japanese ethos') rather than a positional ethic during the modernization process.[71] As Sen describes it, 'If the invisible hand does a great deal of visible good in Japan, the hand does not seem to work through the relentless pursuit of self-interest.'[72] Specifically, what the Japanese model suggests is that *self-interest based objectives* can be more effectively achieved by an alteration in *group* attitudes (to, for example, team work). Indeed:

> when one considers how production takes place in a modern industrial establishment, it is quite incredible to think that being actively self-interested can be such a virtue. Success in production depends greatly on team work, and while that interdependent picture provides incentive for a group, it is not an incentive that can be effectively translated into rewards and punishments related to individual work and performance.[73]

Notes

1. See Lowenthal (1972), pp. 245, 247. Another good example is provided by Beckford's description of plantation economies of the Third World: 'Social aspirations of all groups in plantation society are in the direction of the life style of the planter class. And the demonstration effect of this

metropolitan-oriented group . . . sets the stage, in the society as a whole, for patterns of consumption' (Beckford, 1972, p. 205).
2. Nurkse (1957), p. 65.
3. Hirsch (1976), Scitovsky (1976) and Frank (1985a and b).
4. Nurkse (1957), p. 59.
5. Ibid., pp. 61–2, emphasis added.
6. Stigler and Becker (1977).
7. Hirsch (1976), p. 52.
8. Ibid., p. 52, emphasis in original.
9. Frank (1985a), p. 101.
10. Ibid.
11. Veblen (1899), p. 103.
12. Ibid., p. 112.
13. Ibid., emphasis added.
14. Ibid., p. 98.
15. Ibid., p. 103.
16. A very interesting recent example of the process described by Hirsch is the decline in prestige that is said to have been experienced by wearers of a Rolex watch. Whereas in the 1980s this watch was a signal that one 'had arrived', subsequently 'something has happened to take the shine off the Rolex: too many other people wear them.' 'Wristwatching', *The European*, Elan, 10–12 May 1991, p. 22.
17. Inkeles and Smith (1974), pp. 290–1.
18. Inkeles (1983), p. 178.
19. Ibid., p. 182.
20. Ibid., p. 19.
21. Ibid., p. 20.
22. Ibid., p. 54.
23. Ibid., p. 55, emphasis in original.
24. Ibid., p. 56.
25. Ibid., p. 318.
26. Portes (1973), p. 270.
27. It is essential to recognize that Nurkse's view does *not* necessarily mean that the individual consumer is bereft of any sort of social influence on his consumption behaviour. His relationship to others may instead be one in which he *communicates* his modernity to them by consuming goods that are imported from developed countries. This wider view of 'consumption as communication' is very much in line with what Douglas and Isherwood propose. For them, in fact, Veblen's analysis of conspicuous consumption 'has much to answer for when we consider how widely his analysis of the leisure class is received' (Douglas and Isherwood, 1979, p. 4). They advocate an alternative approach to consumption in which 'Goods assembled together in ownership make physical, visible statements about the hierarchy of values to which their chooser subscribes . . . goods in their assemblage present a set of meanings, more or less coherent, more or less intentional. They are read by those who know the code and scan them for information. The great novelists have never doubted just how far removed this function of creating meanings is from the uses of goods for welfare and display' (Douglas and Isherwood, 1979, p. 5).

Csikszentmihalyi and Rochberg-Halton (1981) make a similar point by distinguishing between two meanings that goods convey. On the one hand, goods serve as a means of individual *differentiation*: that is, a means of separating the owner from the social context by emphasizing his individuality. This role of goods, they contend, has been greatly overemphasized. 'Thus although status symbols are an extremely important aspect of the whole person–object interaction process, it is, nevertheless unfortunate that this one dimension has so overshadowed the rest that it is almost impossible to think of people's possessions except as symbols of their social standing' (Csikszentmihalyi and Rochberg-Halton, 1981, p. 31). The second (and less emphasized) meaning conveyed by goods, in contrast, involves social *integration*: 'They might represent dimensions of similarity between the owner and others: shared descent, religion, ethnic origin, or life-style' (Czikszentmihalyi and Rochberg-Halton, 1981, pp. 40–1). In this case, the good expresses the belonging of the owner to the social context rather than his differentiation from it.

Other important references on the communicative aspects of consumption include Leach (1976), Appadurai (1986), Solomon (1983), Mukerji (1983), Sahlins (1976), Goffman (1956), Holbrook and Hirschman (1980), McCracken (1988), and from an economics perspective, Spence (1974).

28. Portes (1973), pp. 270–1, emphasis added.
29. Yet another possible view – associated with Berger, Berger and Kellner (1974) – attempts to combine the general and specific modes of preference formation in developing countries. These authors argue that there are certain 'intrinsic' relations between modern technological production and modern consciousness such as the sense of 'componentiality' that is imparted to the factory worker. 'This apprehension of reality in terms of components is essential to the reproducibility of the production process as well as to the correlation of men and machines' (Berger, Berger and Kellner, 1974, p. 32). In addition, Berger, Berger and Kellner identify a set of 'extrinsic' processes which are linked to technological production more loosely through a variety of historical processes and which, accordingly, will vary from one case to another. (In capitalist, but not socialist, countries, for example, there are extrinsic elements of technological production which derive specifically from 'the motives of individual competition and profit-making geared to the accumulation of private property'.) In general, the authors conclude that what 'manifests itself empirically as modern consciousness is a highly complex combination of these "necessary" and "accidental" elements' (p. 90).
30. Maslow (1954).
31. Though as we indicate below the Maslow hypothesis does not always seem to stand up to empirical scrutiny.
32. Wilson (1972), p. 14.
33. Ibid., p. 94.
34. Ibid., p. 226. Another interesting anthropological study of the Caribbean region, by Lieber (1981), suggests that positional behaviour (in urban Trinidad) is apparent with respect to other dimensions of status than affluence alone. Because it 'eludes precise definition', ethnicity, for

example, 'involves the active use of cultural signifiers, or clusters of symbols which, serve as vehicles for marking a person's (or group's) social style and position.' These serve in fact as 'resources that the ethnically ambiguous actor may attempt to manipulate to his own advantage' (Lieber, 1981, p. 104). Similarly, 'Since jobs are related to class status, job seekers attempt to secure those sorts of positions which on the surface look good to other people. Hence a job in a modern, air-conditioned office which allows an employee to wear a white shirt is often preferred to what may be a far better and more highly paid job which has a less prestigious ambience. The signs of a position tend to be very conspicuously displayed' (Lieber, 1981, pp. 98–9).

35. Ibid., p. 231.
36. Ibid., p. 223, emphasis added.
37. Sauvant (1976), pp. 68–9.
38. Hirsch (1976), p. 82.
39. Jelliffe (1972), p. 201.
40. Ledogar (1975).
41. Mattelart (1983), p. 38 quotes a suggestion from Business International regarding the sales promotion of Western products in the Third World that contains clear positional implications. The suggestion is to 'try and give your products a Westernized appearance to give them social standing in regions undergoing rapid development wherein ideas of modernization and Westernization are linked.'
42. Mitchell and Epstein (1959), p. 32, emphasis added.
43. Veblen (1899), p. 87. These are not, however, the only reasons why consumption in urban areas may be more prone to positionality. One needs to consider in addition the significance of the fact that in traditional (rural) societies rank may be almost entirely excluded, or it may be 'ascribed' (Smith, 1966). In the latter case, rankings are ascribed to persons at birth and cannot subsequently be altered. The best-known case in point is the Indian caste system under which, 'virtually every aspect of the individual's future life was determined by his birth into a particular caste' (Smelser and Lipset, 1966, p. 8). In another traditional form of ranking system, rank does vary but according to certain *exogenous* events. For example, in one of the South Indian villages studied by Scarlett Epstein (1962), 'A man's economic status is determined by the size of the ancestral estate and the number of heirs that have to share it; personal initiative can hardly help to raise his economic status. Although the relative economic status of successive generations of households may be quite different, this has nothing to do with the personal ability and efforts of the particular householder, but rather is a result of the accidents of births and deaths' (Epstein, 1962, p. 324). Other examples are provided in Smith (1966).

    Both ascriptive ranking systems and traditional systems in which rank varies exogenously contrast with what may be termed 'achievement-oriented' societies, in which rank is variable and depends on some kind of behavioural performance criterion. (This last most closely approximates the situation in industrial societies in the West, but it is also found in some traditional societies, as Mead (1961), for example, has shown.)

In general one may expect positional competition to be most intense in achievement-oriented societies and to be less evident where rank is ascribed or where it is exogenously determined. For it is only when the possibility of altering one's rank by behavioural means exists that the scope for such competition would seem to arise. (A comparative study of 13 traditional societies with different types of ranking systems provides some indirect support for this proposition (Mead, 1961).) Studies of the urbanization process in particular developing economies are also suggestive in this regard. A.L. Epstein's observation of the African situation, for example, notes that 'the marked emphasis on achieved as against ascribed status', among other factors produces 'a set of conditions in which the struggle for prestige becomes a major preoccupation of the new urban dwellers' (Epstein, 1969, p. 105).

Yet, even in ascriptive societies one still frequently encounters consumption with an apparently positional component. In the Indian caste system, for example, there appears to be intense competition between members of a *given* caste, which takes place in a context where the ritualistic ranking system coexists with a ranking that is based on economic factors (Epstein, 1962). Furthermore, conspicuous consumption in ascriptive societies may be undertaken in order to confirm or validate one's rank rather than to advance positionally. Indeed, according to some authorities, even the famous potlatch ceremony of the Southern Kwakiutl can be viewed as a device to validate high status (of the chiefs) rather than to 'give, or create, social status' (Drucker and Heizer, 1979, p. 268).

There is a very large literature on these and other aspects of status seeking in traditional societies to which the interested reader is referred. (See, for example, Mead (1961), Douglas and Isherwood (1979), Suttles (1979), Piddocke (1979), Drucker and Heizer (1979), Smith (1966), Epstein (1962), Smelser and Lipset (1966), Appadurai (1986), Marriott (1968), Herskovits (1952), Dalton (1971), Mason (1981), Plotnicov and Tuden (1970), Bohannan and Dalton (1962).) Some of these contributions also compare status-seeking consumption behaviour in traditional versus modern societies. Douglas and Isherwood (1979) and Appadurai (1986), for example, see in the attempts by the elite to use consumption as a means of excluding others a factor that has universal validity. Thus, 'From the point of view of demand, the critical difference between modern, capitalist societies and those based on simpler forms of technology and labor is *not* that we have a thoroughly commoditized economy whereas theirs is one in which subsistence is dominant . . . but rather that the consumption demands of persons in our own society are regulated by high-turnover criteria of "appropriateness" (fashion), in contrast to the less frequent shifts in more directly regulated sumptuary or customary systems. In both cases, however, demand is a socially regulated and generated impulse, not an artifact of individual whims or needs' (Appadurai, 1986, p. 32).

44. Veblen (1899), p. 87. It might be possible to test Veblen's hypothesis by comparing the consumption behaviour of urban and rural residents at given income levels. Following Frank's (1985b) suggestion, one could,

more specifically, examine whether spending by the former tends to have a higher 'observable' and a lower 'non-observable' component (since 'conspicuousness' is a necessary requirement of positional consumption). What fragmentary pieces of evidence are available seem to be consistent with this expectation. Consider first the 'unobservables', two of the most important of which are savings and (within limits) nutritional intake. With respect to the former, Gillis *et al.* (1983) find that 'There is a strong tendency around the world for rural households to save higher fractions of their incomes than urban households *at comparable levels of income*' (Gillis *et al.*, 1983, p. 282, emphasis added). With regard to nutrition, 'It seems that *for a given income group*, the urban population has a significantly lower nutritional status as measured by caloric intake' (Austin *et al.*, 1976, p. 340, emphasis added). Corresponding to, and indeed partly accounting for these findings, is the tendency for urban consumers to spend more than their rural counterparts with comparable incomes on relatively expensive and observable symbols of status. Such a tendency has been noted, for example, in the case of infant formula (Landgren-Gudina, 1984), crystal sugar (James, 1977), sophisticated soap (Mubin and Forsyth, 1984) and imported detergents (James, 1983). But because they generally control only for income, these studies still leave open the question of whether other variables (such as product access and information) may account at least partly for what appears to be more positional consumption behaviour in urban areas.

45. Veblen (1899), pp. 87–8, emphasis added.
46. Epstein (1962).
47. Ibid., p. 329.
48. Ibid., p. 330–1.
49. We should emphasize that our discussion here follows the Veblen/Hirsch view of status-seeking behaviour; other models of this behaviour, as Congleton (1989) has demonstrated, have different (and sometimes positive) welfare effects. (Favourable externalities, for example, may accrue to those who are not actively involved in status-seeking games.) Moreover, even the Veblen/Hirsch model may be associated with some offsetting welfare effects (such as, for example, on the supply of labour) which we do not address.
50. Veblen (1899), pp. 96–7.
51. Ibid., pp. 154–5.
52. Stewart (1977).
53. Ironmonger (1972).
54. Supple (1981), p. 137.
55. Ibid.
56. MacEwan (1974), p. 212, emphasis added.
57. Southall (1961), p. 21.
58. Grossman and Shapiro (1988).
59. Veblen (1899), p. 85.
60. Ibid., p. 168, emphasis added.
61. Frank (1985b), pp. 155–6. Hirsch, in contrast, takes the more Maslowian view that among the poor consumption is a purely private phenomenon. As he puts it, 'To a hungry man, the satisfaction derived from a square

meal is unaffected by the meals other people eat or, if he is hungry enough, by anything else they do. *His meal is an entirely individual affair*' (Hirsch, 1976, p. 2, emphasis added). Since low-income consumption of this kind is said to be devoid of social content, it necessarily cannot contain any positional elements (since these are merely a special case of social consumption). It is worth noting here that the Maslow hypothesis is not always well supported by the available data; poor consumers do, for example, seem to consume luxury positional goods (such as infant formula) before their more basic needs have been met. For critiques of Maslow see Lederer (1980) and Munro (1983).

62. Ledogar (1975).
63. In urban Upper Volta, for example, it has been observed that 'many women cultivators are impressed with the fact that high-status women, whose babies are considered by all the town's women to be the best cared for, use the bottle. (A corollary of this . . . is that low-status women use preserved milk because by so doing they feel and become *un peu civilisée.*)' See Skinner (1974), p. 189.
64. Austin *et al.* (1976), p. 98. In Brazil a decline over time in nutritional status among the urban poor has been associated with the substitution for food expenditure of durable goods consumption (Wells, 1977).
65. Leibenstein (1975), p. 5.
66. Ibid., p. 6.
67. It could be argued against this, however, that when there is equality and it is known, status seeking would not be very effective since individuals would realize this was just a different allocation of the same income. If, on the other hand, there are marked and known income differences, there may not be much point either. Conceivably, then, it is where differentials are not widely known and not vast that most such behaviour takes place. (I owe this point to Frances Stewart, made in comments on an earlier version of this chapter.)
68. McClelland (1976). It is interesting in this regard to note that the predominant reason often given by individual urban women for weaning in formal interviews (viz. milk insufficiency) differs from the motivations that are elicited during more informal discussions (Austin *et al.*, 1976).
69. Douglas and Isherwood (1979), pp. 36–7.
70. For details see Douglas and Isherwood (1979). The cooperative societies described in Mead (1961) also provide concrete illustrations of the various modes by which the group environment tends to constrain positional competition (as reflected in a weak emphasis on rising in status in these societies). And even in highly individualistic advanced economies such as the United States, the group, in the form of the trade union, is said by Frank (1985) to implement collective agreements to protect 'nonobservables' (such as savings and insurance). Policy too could seek to increase group influence on decisions, by, for example, reducing private control over consumption and increasing public (as for instance by state-provided goods and services). In addition, one might, as Layard (1980) has suggested, attempt to discourage positional as opposed to other forms of gaining status, by, for example, 'getting adults to spend less time comparing the performances of young children' (p. 749) and by granting

honours, titles and other kinds of acknowledgments of diverse forms of achievement. Still other policies could be directed towards the status symbols themselves: according to our analysis, for example, there is a rationale for tariff protection against the import of products which are intensive in positional characteristics and low in their embodiment of essential characteristics. In some extreme cases (such as that of infant formula) these inappropriate symbols of status might be banned altogether, thus removing them from the relevant areas of choice.

71. See, for example, Lockwood (1954) who notes that 'By comparison with the value standards prevailing among many other peoples, perhaps most of them, social position and prestige in Japan are less dependent on display of material wealth' (p. 284).
72. Sen (1984), p. 104.
73. Ibid.

References

Appadurai, A. (1986). *The Social Life of Things*, Cambridge University Press.

Austin, J.E. *et al.* (1976). *Urban Malnutrition*, a report submitted to the World Bank by Harvard University.

Beckford, G. (1972). *Persistent Poverty*, Oxford University Press.

Berger, P., Berger, B. and Kellner, H. (1974). *The Homeless Mind*, Penguin.

Bohannan, P. and Dalton, G. (eds) (1962). *Markets in Africa*, Northwestern University Press.

Congleton, R. (1989). 'Efficient Status Seeking: Externalities and the Evolution of Status Games', *Journal of Economic Behavior and Organization*, Vol. 11.

Csikszentmihalyi, M. and Rochberg-Halton, E. (1981). *The Meaning of Things*, Cambridge University Press.

Dalton, G. (1971). *Economic Anthropology and Development: Essays on Tribal and Peasant Economies*, Basic Books.

Douglas, M. and Isherwood, B. (1979). *The World of Goods*, Allen Lane.

Drucker, P. and Heizer, R. (1979). 'To Make My Name Good: A Reexamination of the Southern Kwakiutl Potlatch', in McCurdy, D. and Spradley, J. (eds), *Issues in Cultural Anthropology*, Little, Brown.

Epstein, A.L. (1969). 'The Network and Urban Social Organization', in Mitchell, J.C. (ed.), *Social Networks in Urban Situations*, Manchester University Press.

Epstein, T.S. (1962). *Economic Development and Social Change in South India*, Manchester University Press.

Frank, R. (1985a). 'The Demand for Unobservable and Other Nonpositional Goods', *American Economic Review*, March.

Frank, R. (1985b). *Choosing the Right Pond: Human Behavior and the Quest for Status*, Oxford University Press.

Gillis, M., Perkins, D., Roemer, M. and Snodgrass, D. (1983). *Economics of Development*, Norton.

Goffman, E. (1956). *The Presentation of Self in Everyday Life*, University of Edinburgh.

Grossman, G. and Shapiro, C. (1988). 'Foreign Counterfeiting of Status Goods', *Quarterly Journal of Economics*, Feb.

Herskovits, M. (1952). *Economic Anthropology*, W.W. Norton.

Hirsch, F. (1976). *The Social Limits to Growth*, Harvard University Press.

Holbrook, M. and Hirschman, E. (1980). 'Symbolic Consumer Behavior: An Introduction', *Proceedings of the Conference on Consumer Esthetics and Symbolic Consumption*, New York University, 16 and 17 May.

Inkeles, A. (1983). *Exploring Individual Modernity*, Columbia University Press.

Inkeles, A. and Smith, D. (1974). *Becoming Modern*, Harvard University Press.

Ironmonger, D. (1972). *New Commodities*, Cambridge University Press.

James, J. (1977). *Technology, Products and Income Distribution: A Conceptualisation and Application to Sugar Processing in India*, ILO, World Employment Programme Research, Working Paper No. 61.

James, J. (1983). *Consumer Choice in the Third World*, Macmillan.

Jelliffe, D. (1972). 'Commerciogenic Malnutrition', *Nutrition Review*, Sept.

Landgren-Gudina, M. (1984). 'Weaning Food and Low-Income Consumers in Ethiopia', in van Ginneken, W. and Baron, C. (eds), *Appropriate Products, Employment and Technology: Case Studies on Consumer Choice and Basic Needs in Developing Countries*, Macmillan.

Layard, R. (1980). 'Human Statisfaction and Public Policy', *Economic Journal*, Vol. 90, Dec.

Leach, E. (1976). *Culture and Communication: the Logic By Which Symbols Are Connected*, Cambridge University Press.

Lederer, K. (ed.) (1980). *Human Needs*, Oelgeschlager, Gunn and Hain.

Ledogar, R. (1975). *Hungry for Profits*, IDOC.

Leibenstein, H. (1975). 'The Economic Theory of Fertility Decline', *Quarterly Journal of Economics*, Feb.

Lieber, M. (1981). *Street-Life: Afro-American Culture in Urban Trinidad*, Schenkman.

Lockwood, W.W. (1954). *The Economic Development of Japan*.

Lowenthal, D. (1972). *West Indian Societies*, Oxford University Press.

McClelland, D. (1976). 'Review of A. Inkeles and D. Smith, *Becoming Modern*', *Economic Development and Cultural Change*, Oct.

McCracken, G. (1988). *Culture and Consumption*, Indiana University Press.

MacEwan, A. (1974). 'Differentiation Among the Urban Poor: An Argentine Study', in de Kadt, E. and Williams, G. (eds), *Sociology and Development*, Tavistock.

Marriott, M. (1968). 'Caste-Ranking and Food Transactions', in Singer, M. and Cohn, B. (eds), *Structure and Change in Indian Society*, Aldine.

Maslow, A. (1954). *Motivation and Personality*, Harper & Row.

Mason, R. (1981). *Conspicuous Consumption*, Gower.

Mattelart, A. (1983). *Transnationals and the Third World*, Bergian & Garvey.

Mead, M. (1961). *Cooperation and Competition Among Primitive Peoples*, 2nd edn, Beacon Press.
Mitchell, J.C. and Epstein, A.L. (1959). 'Occupational Prestige and Social Status Among Urban Africans in N. Rhodesia', *Africa*, No. 29.
Mubin, A. and Forsyth, D. (1984). 'Technology, Employment and Income Distribution: the Soap Industry in Bangladesh', in van Ginneken, W. and Baron, C. (eds), *Appropriate Products, Employment and Technology: Case Studies on Consumer Choice and Basic Needs*, Macmillan.
Mukerji, C. (1983). *From Graven Images: Patterns of Modern Materialism*, Columbia University Press.
Munro, D. (1983). 'Developing "*La Conscience Professionelle*" in Cultures with a History of Vocational Disadvantage', in Blackler, F. (ed.), *Social Psychology in Developing Countries*, Wiley.
Nurkse, R. (1957). *Problems of Capital Formation in Underdeveloped Countries*, Blackwell.
Piddocke, S. (1979). 'The Potlatch System of the Southern Kwakiutl: A New Perspective', in McCurdy, D. and Spradley, J. (eds), *Issues in Cultural Anthropology*, Little, Brown.
Plotnicov, L. and Tuden, A. (1970). *Essays in Comparative Social Stratification*, University of Pittsburgh Press.
Portes, A. (1973). 'Modernity and Development: A Critique', *Studies in Comparative International Development*, Vol. 8 (3), Fall.
Sahlins, M. (1976). *Culture and Practical Reason*, University of Chicago Press.
Sauvant, K. (1976). 'The Potential of Multinational Enterprises as Vehicles for the Transmission of Business Culture', in Sauvant, K. and Lavipour, F. (eds), *Controlling Multinational Enterprises*, Westview.
Scitovsky, T. (1976). *The Joyless Economy*, Oxford University Press.
Sen, A. (1984). *Values, Resources and Development*, Harvard University Press.
Skinner, E.P. (1974). *African Urban Life: The Transformation of Ouagadougou*, Princeton University Press.
Smelser, N.J. and Lipset, S.M. (eds) (1966). *Social Structure and Mobility in Development*, Aldine.
Smith, M.G. (1966). 'Pre-Industrial Stratification Systems', in Smelser, N.J. and Lipset, S.M. (eds), *Social Structure and Mobility in Development*, Aldine.
Solomon, M. (1983). 'The Role of Products as Social Stimuli: A Symbolic Interactionism Perspective', *Journal of Consumer Research*, Vol. 10, Dec.
Southall, A.W. (1961). 'Introductory Summary', in Southall, A.W. (ed.), *Social Change in Modern Africa*, Oxford University Press.
Spence, A.M. (1974). *Market Signaling: Informational Transfer in Hiring and Related Screening Processes*, Harvard University Press.
Stewart, F. (1977). *Technology and Underdevelopment*, Macmillan.
Stigler, G. and Becker, G. (1977). 'De Gustibus Non Est Disputandum', *American Economic Review*, March.
Supple, B. (1981). 'Income and Demand 1860–1914', in Floud, R. and McCloskey, D. (eds), *The Economic History of Britain Since 1700*, Vol. 2, Cambridge University Press.

Suttles, W. (1979). 'Affinal Ties, Subsistence, and Prestige Among the Coast Salish', in McCurdy, D. and Spradley, J. (eds), *Issues in Cultural Anthropology*, Little, Brown.
Veblen, T. (1899). *The Theory of the Leisure Class*, Macmillan.
Wells, J. (1977). 'The Diffusion of Durables in Brazil and Its Implications for Recent Controversies Concerning Brazilian Development', *Cambridge Journal of Economics*, Vol. 1 (3).
Wilson, P. (1972). *Crab Antics – The Social Anthropology of English Speaking Negro Societies of the Caribbean*, Yale University Press.

7 The Sources of Frustration and Disappointment with the Introduction of New Products in Developing Countries

The notion of positional characteristics that was introduced in the previous chapter led us to suggest that the satisfaction derived from certain new products may in fact turn out to be less than was initially expected of them. The sense of frustration and disappointment that is the likely result stands in stark contrast to the unquestioned benefits that traditional consumption theory tends to associate with new products. But consumption undertaken (partly or entirely) for positional reasons is not the only route by which new products can produce these reactions in the Third World. On the contrary, there are a variety of different ways in which this can occur, as we shall try to demonstrate below.

The main theme of the chapter is that disappointment and frustration take forms that differ according to which particular assumptions of traditional consumption theory are rejected. These differences bear, among other ways, on the nature of the commodities that yield disappointing outcomes and on the types of consumers to which these outcomes apply.

DISAPPOINTMENT, FRUSTRATION AND TRADITIONAL CONSUMPTION THEORY

In traditional consumption theory there is never a discrepancy between the utility that the consumer expects to achieve and the utility that is actually realized. That is, 'There are no surprises with respect to the amount of utility received.'[1]

Perhaps the most obvious assumption that this conclusion requires is perfect knowledge: that is, 'In accordance with this postulate, people are supposed to calibrate their purchases and time uses by matching their preferences, which are fully known to them, against the equally well-known world of available consumption experiences.'[2] In addition to this requirement that preferences be known, the 'absence of surprises' in traditional theory also demands that preferences be constant. For once change occurs, it always carries with it the possibility that the individual judges himself to be worse off in the new situation than in the old.

Another assumption that has the effect of excluding disappointment from the purview of traditional theory is that individual satisfaction depends only on one's own consumption. This assumption excludes not only the positional-type problems that were described in the previous chapter, but also and more generally what economists would describe as 'negative consumption externalities'.

Yet another means by which expected and actual utility are always equated is the assumption implicit in traditional theory that consumption occurs at the point where commodities are purchased. This assumption effectively excludes all the factors that make the characteristics derived from commodities in general and new products in particular a highly variable magnitude, rather than an amount that is the same for all consumers (of a given product). In other words, by allowing no discrepancy to exist between the potential and the realized characteristics of new products, the theory permits no disappointment to arise from the way in which these products are actually used.

In what follows, by systematically reviewing each of these assumptions, an attempt is made to derive and to classify the various sources of disappointment that are associated with new products in developing countries.

INFORMATION IMPERFECTIONS

We noted in Chapter 2 that advertising is systematically associated with the introduction of new products in developing countries. As was also suggested there, the intensity of this advertising may itself lead to an overestimation of the benefits of new products. To some degree, this may reflect the view that it is only successful and worthwhile brands that are able to be advertised and its corollary that less (or un-) advertised brands are inferior. Any such tendency (which will be especially pronounced in the case of products such as pharmaceuticals, detergents and cigarettes that are often the most heavily advertised) will be reinforced by the frequently observed propensity of consumers in developing countries to regard goods imported from the developed countries as superior to locally produced substitutes. In so far as the actual experience with new products contradicts the exaggerated expectations that these factors tend to engender, a sense of disappointment may often (though, as indicated below, not necessarily) be the result.

Disappointment may of course also be related to the *nature* of the advertising that is associated with new products. In particular, as was suggested in Chapter 4, the information that is conveyed may often be such as to cause the consumer to overestimate the extent to which these products contain valued characteristics and to underestimate the extent to which they possess unwanted characteristics. In some cases, the disappointment that results from these information imperfections may not be very significant and indeed may be offset, partly or fully, by the well-established tendency for consumers to rationalize the wisdom of their choices (according to the psychological theory of cognitive dissonance). This is essentially what James found in his case study of the effects of advertising in the market for laundry cleaning products in the Caribbean.[3] In other cases, however, disappointment may be much more pronounced and less susceptible to the offsetting psychological forces exerted by cognitive dissonance. (For, as Hirschman has rightly emphasized, it is difficult to imagine consumers trying to convince themselves of the wisdom of grossly disappointing choices.)[4]

There is now substantial evidence, for example, that multinationals and other firms in the pharmaceutical industry have made vastly exaggerated claims about the efficacy of certain drugs; there is evidence too to suggest that warnings about the potentially adverse effects of these products have not always been adequately disclosed.[5] It does not require a great leap of the imagination to believe that these informational deficiencies have accounted, at least in part, for some of the equally well documented cases in which consumers have made obviously inappropriate, and sometimes dangerous, choices of medicines. Consider, by way of example, how the inappropriate selection of antibiotics has created severe problems of drug resistance in certain parts of the Third World. Two of these drugs in particular (streptomycin and penicillin) are highly effective in treating a range of bacterial infections.

> But if these antibiotics are over-used the bacteria they attack quickly build up resistance. Bacteria can make themselves resistant to streptomycin very fast, so the drug becomes useless for TB

sufferers. They then end up paying for more costly *second-line* antibiotics. Drug resistance presents problems everywhere. But it can make the difference between life and death in poor countries. . . . Despite the obvious dangers, in many Third World countries manufacturers are marketing products containing streptomycin and other first-line antituberculous drugs for trivial, unnecessary uses. For example, UNCTAD reports that many cough and cold remedies containing streptomycin are sold in Nepal, a country where almost a quarter of child deaths are caused by TB.[6]

The infant formula case is also relevant in this regard, for it is often alleged that misleading advertising and marketing have caused a decline in breastfeeding, especially but not only in urban areas of developing countries.[7] And the highly adverse consequences that this has often had for the health of infants have been amply described in a number of the preceding chapters.

It would be inaccurate, however, to ascribe these and other cases where disappointment has followed the consumption of new products entirely to advertising and marketing practices. For what also needs to be recognized is that governments in many developing countries have tended to do little or nothing either to regulate these practices or to provide consumers with the information that would have permitted them to make more informed choices. This lack of a countervailing state influence is perhaps most clearly illustrated in the case of cigarette advertising. In particular:

In most developing nations, the smoking/health issue has been almost totally absent from public debate. The adverse health effects of cigarette smoking have not become public knowledge nor the object of governmental action, with the consequence that TNCs often find it much easier to market their products in these areas than at home . . . A World Health Organization (WHO) survey showed that of 100 governments providing information, 70 had no legislation whatsoever aimed at controlling the promotion or use of cigarettes . . . Another survey revealed that over two dozen LDCs (including all 14 Latin American nations surveyed) had taken no action in the regulation of cigarette advertising and the great majority had no warnings on cigarette packs.[8]

It also needs to be recognized that even when accurate information is available to the consumer, it is not always utilized. There is a lot of evidence, for example, that information about the correct usage of powerful modern drugs is not used by a remarkably high percentage of patients. Some of this evidence is presented below in the section dealing with discrepancies between what is potentially obtainable from the consumption of new products and what is actually realized.

EXTERNALITIES IN CONSUMPTION

In the form of environmental pollution, negative externalities associated with the consumption of modern products (such as automobiles, household sprays, paints and pharmaceuticals) have received considerable attention especially in recent years. Understandably, much of this attention has focused on the developed countries, where these products are usually most prevalent.

Yet, particularly in some of the more industrialized of the developing countries, similar problems have already emerged. Indeed, in at least one such case, that of Mexico City, these problems appear to be even more pronounced than in the developed world. In particular, ozone pollution in that densely inhabited location is said to be the worst in the world and much of it is thought to be due to toxic emissions from automobiles.[9]

More generally, there is a fear that in the future, 'As the poorer countries industrialise, buy cars, get richer, their capacity for [environmental] damage will overtake that of the rich world because they have more people.'[10]

Whether they take this form (of environmental pollution) or, as in Chapter 6, of positional goods, externalities of consumption undermine the assumption that individual welfare increases in simple proportion to the consumption of new goods. Instead, in both cases, what seem *initially* to be gains of this kind, turn out subsequently to be in some degree illusory; gains for the individual cannot be sustained collectively. Despite this fundamental similarity, however, there are also some important differences in the nature of the disappointment that is associated with positional consumption on the one hand and with 'ordinary' (negative) externalities of consumption on the other.

For one thing, whereas the former is extremely difficult to measure, many forms of environmental damage can be assessed with relative accuracy. A second distinction is that whereas externalities of the positional type may affect only those involved in 'playing the game' (of seeking to enhance status through strategic consumption choices), environmental damages have the capacity to afflict all the inhabitants of a region and in some cases even those who live in other regions. (Pollution due to automobile emissions, for example, affects both those who own automobiles and those who do not.) Yet another distinction has to do with the probable time lag that elapses between the initially favourable welfare effects of consuming new products and the subsequent disappointment to which they (collectively) give rise. For, whereas positional goods may often be associated with the rather fleeting changes of fashion that are related to and indeed encouraged by, the advertising and marketing efforts of producers of, say, clothing or automobiles, the environmental consequences of consuming certain new goods may take a much longer time to materialize.[11]

In some cases, of course, new goods may be subject to both these difficulties: one that occurs in the relatively short run as the positional characteristics of the new product quickly elude the consumer (of, say, a new automobile) and one that occurs in the longer run as the subsequent environmental damages become apparent through widespread use of the product.

PREFERENCE CHANGE

Under this heading fall two distinct sources of disappointment with new products that operate through very different causal mechanisms and which tend to affect very different groups of consumers in developing countries. In some cases, moreover, the two sources of disappointment apply to different product categories. Consider first the source of disappointment that is associated with the welfare economics of variable tastes.

The welfare economics of variable tastes

Based as it is on ordinal utility theory (or indifference curve analysis), traditional welfare economics has no way of dealing with the consequences of preference change. For there is then no means of judging whether his old or his new tastes should be chosen as the relevant yardstick. One way out of this difficulty has been described by Harsanyi as the 'quasi-cardinal' approach,[12] and for reasons that will become apparent below, it applies particularly well to the type of

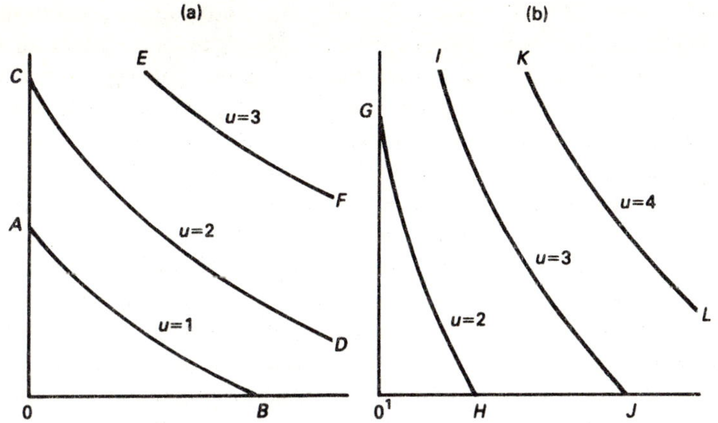

Figure 7.1 The quasi-cardinal approach.
Source: Harsanyi (1954), Fig. 1.

preference change with which we are here especially concerned.

The key assumption of this approach is that:

> [Consumers] are able to compare the utilities they would derive from different groups of commodities at a given moment (on the basis of given taste), but also that they are able to compare the utility (subjective satisfaction) they derived from one group of commodities at the time t_1, when their taste was T_1, with the utility (or subjective satisfaction) they now derive from another group of commodities at the time t_2, when their taste is T_2.[13]

Though it may seem to, this assumption does not in fact depend on cardinal utility measurement – that is, the measurement of differences between levels of utility. What it does require is somewhat less demanding, namely the assumption that 'if certain objective situations have the same value for the individual on the basis of his present taste T_2 as certain other situations had on the basis of his previous taste T_1, they must be assigned the same value of the utility function as the value ascribed to the latter situations.'[14]

Exactly what role this assumption plays in welfare comparisons with variable tastes, can most easily be shown in terms of the simple diagram given in Figure 7.1. Figures 7.1(a) and (b) represent the consumer's indifference map corresponding to his initial and his new tastes respectively. Ordinal utility analysis of this change tells us only that higher indifference curves in each map represent greater levels of satisfaction for the consumer; it thus provides no way of *comparing* the utility levels associated with the indifference curves of the two maps. In short, we have no way of making welfare comparisons of the two situations. At the other extreme, the assumption that utilities are cardinally measurable would, of course, be able to ascribe absolute utility levels to each indifference curve belonging to both maps (which curves, accordingly, become directly comparable).[15] The 'quasi-cardinal' method falls between these two forms of measurement in that it requires any indifference curve in Figure 7.1(b) that has an equivalent in Figure 7.1(a) to be assigned the same absolute utility value as the latter. 'This piece of information may be called the *correspondence rule* between the two indifference maps.'[16]

For example, the indifference curves in Figure 7.1(a) have been assigned the arbitrary values $u = 1$, $u = 2$ and $u = 3$. Two of the curves in Figure 7.1(b) (*GH* and *IJ*) correspond in this sense to the curves *CD* and *EF* in Figure 7.1(a). *GH* and *IJ*, accordingly, are assigned the same values as *CD* and *EF* respectively.

> On the other hand, indifference curves of the second graph [(b)] that have no equivalent on the first graph – in so far as they represent a utility lower, or higher, than any point of the first graph [(a)] – can be assigned an arbitrary value of the utility function, except that this value must be lower than the one corresponding to any preferred position and higher than the one corresponding to any less preferred position. Therefore *KL* has been given an index higher than the index of any indifference curve drawn on the left-hand graph.[17]

Following Harsanyi let us consider next how this approach can be applied to assessing the welfare effects of changes in preferences that are associated with the introduction of new products. This framework, in turn, can then be applied to the specific context of preference changes that are related to the introduction of new products in developing countries.

Figure 7.2 shows, as before, the consumer's indifference map before and after his change in preferences. The difference is that in this case the change is associated with the introduction of a new product, *X*. Money is now measured along the vertical axis and the amount of the new commodity is measured along the horizontal axis. In the initial situation, represented by Figure 7.2(a), the indifference curves

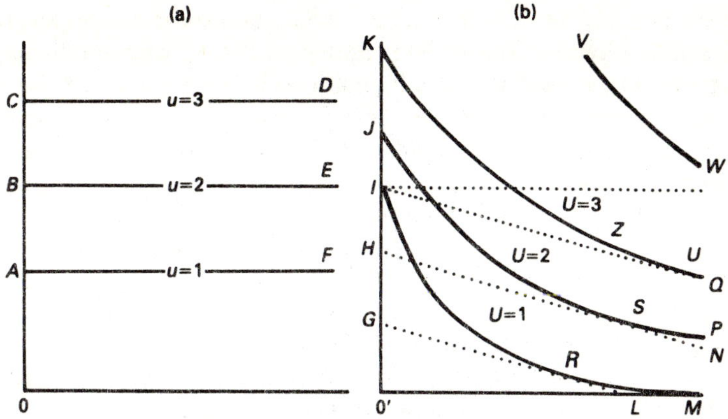

Figure 7.2
Source: Harsanyi (1954), Fig. 2.

are all horizontal by definition, since the consumer has no demand for the new product *X*. The consumer's money income is assumed to be equal to 0*B*, which is equivalent to 0^1I. Whereas the consumer's initial equilibrium is represented by point *B* in Figure 7.1(a), in the new situation he moves to point *U*, where the budget line is tangential to the highest indifference curve available to him (*KZ*). Whether the new situation represents a position of higher or lower utility than the initial one depends, of course, 'on the correspondence rule' between the two indifference maps and 'in particular, on which indifference curve of the second map is equivalent to the line BE of the first map'.[18]

While he identifies three possible cases, Harsanyi suggests that the most common in practice is one in which the indifference curve equivalent to *BE* lies in between *IM* and *VW*.

> In this case *U* is a position preferred to *B* if it lies above the curve equivalent to *BE*, is a position less preferred than *B* if it lies below that curve, and is equally preferred to *B* if it lies exactly on that curve. Since the position of *U* depends on the slope of the price line *IQ*, this implies that *U* is a position preferred to *B* if the price of *X* remains below a certain limit and a position less preferred than *B* if the price of *X* is higher than that. In other words . . . the consumer enjoys his new taste *if the price of X permits him to gratify his new taste to a sufficient extent, but dislikes it in the opposite case.*[19]

In an entirely analogous way, a similar conclusion is reached with regard to income. Specifically, 'change in a consumer's indifference map would . . . benefit him if his income is high, and would harm him if his income is low, and *would express a shift in his taste away from simpler and less expensive commodities towards more refined and more expensive ones*.'[20]

What light does this framework throw on the welfare effects of preference changes that are associated with the introduction of new products in developing countries and those changes in particular that give rise to a sense of frustration and disappointment? To begin with, the key assumption of this approach – that utility comparisons can be made by the consumer prior and subsequent to his preference change – probably applies best to what in earlier chapters were referred to as changes in values (or, in Hirschman's terms, 'non-wanton' changes in preferences). For it is more likely that the consumer whose preferences undergo so fundamental a change, will have a well-defined view of the desirability of the new situation than the consumer whose preference change conforms merely to the 'wanton' (apples vs. pears) variety.[21]

For this more fundamental type of preference change, the Harsanyi framework provides an important link between the notion of appropriate products that was introduced in Chapter 2, and a form of disappointment that is probably particularly prevalent in developing countries where modernizing attitudes (defined as in the previous chapter) have been relatively widely diffused among the population, but where (as in, say, Brazil) affluence remains the preserve of only a small elite minority. For the low-income inhabitants of such countries – who have acquired the modern values and preferences described in the previous chapter – will be confronted (according to Chapters 2 and 3) with products whose high-income characteristics generally make them appropriate for high rather than low-income groups.[22] The probable outcome, according to the framework we have just described, is that the consumer's inability to indulge his preferences for these products will tend to leave him with a sense of frustration and disappointment. (Of course, it is possible that the low-income consumer will purchase the high-income product, but as noted in Chapter 2, this would cause an imbalance in his consumption pattern.)

The widespread use of infant formula in many parts of the Third World is perhaps the example that is most suggestive of this form of disappointment. For on the one hand, there is evidence that even mothers with very low incomes choose formula in place of breastfeeding of their infants.[23] On the other hand, because the former method of feeding is far too expensive for most such households, it has to be diluted and otherwise improperly used, often with the dire consequences that were described earlier. At the same time, however, the rejected alternative of breastfeeding becomes increasingly unavailable to the mother of the bottle-fed infant.

More generally, the difficulty of 'indulging' the preferences of consumers with low incomes for many new products arises not merely from the basic design of these products for consumers in the rich countries. To this problem have to be added the price-increasing tendencies associated with the import substitution policies that were pursued by so many governments in the post-war period. Protected as they typically were by high rates of effective protection, multinationals and (mostly large) domestically owned firms were able to produce (often inferior) local substitutes for imports of consumer goods at what often were exorbitantly high prices.

Consider, for example, what occurred in Kenya. A World Bank Report on that country published in 1983, noted that it was 'possible to document high local prices and/or low or deteriorating quality for virtually the complete range of products for which there are import restrictions'.[24] Among the most striking examples cited by the Report are the following. 'While importers state that a good quality imported pen could be sold at half a shilling, the domestically produced item sells for KSh [Kenya shillings] 1/75. Toothbrush production is also protected . . . The cheapest line sells for KSh 18 per dozen. Traders report they could readily import equivalents at KSh 5 per dozen.'[25]

Kornai has used precisely these sorts of examples to challenge the notion of 'consumer sovereignty' that is so central to traditional consumption theory and to orthodox welfare economics.[26] He distinguishes between the concepts of 'pressure' (which refers essentially to the conditions prevailing in a buyers' market) and 'suction' (which refers to what is often described as a sellers' market). Only where the former applies, so Kornai's argument goes, is the seller '*compelled to adapt to the buyers' requirements; it is only in that case that the consumer is "sovereign"*. (And even in this case he is sovereign only in the short run. In actual fact his needs are influenced by technical progress and by the new products appearing in the market.)'[27] Where, in contrast, it is suction that applies over a wide range of economic activity, 'the buyers will compete for the seller and the latter will "reign" even in the branches where production and selling are atomized.'[28]

To the sale of high-income products in markets that are frequently best described by 'suction' rather than 'pressure', one needs to add still another important factor which may limit the extent to which the low-income consumer in a developing country is able to indulge his new tastes. This factor is based on the recognition that it may often be inadequate to analyse new products in isolation from one another. In particular, these products may instead be highly complementary not only in the technical sense (that, for example, consumer goods require batteries and cars need roads), but also from a psychological or sociological point of view.

As anthropologists often emphasize, goods in general tend to 'go together' in this latter sense for a number of reasons. One of them derives from the 'advertising and fashion system' which constantly and deliberately influences our view of 'what things go together'.[29] Another, more general, reason is to be found in the role that consumption plays in the communication of material culture. This role comprises far more than just the Hirsch/Veblen-type effects which economists have tended to emphasize. Anthropologists, in contrast, take the much broader view that consumption communicates a multiplicity of messages to others, which include, but are by no means confined to, messages about the superiority of one's rank and position. According to McCracken, for example:

> It appears to be the case that consumer goods do not communicate well when they exist in isolation or in heterogeneous groups. The meaning of a good is best (and sometimes only) communicated when this good is surrounded by a complement of goods that carry the same significance . . . In other words, the symbolic properties of material culture are such that things must mean together if they are to mean at all. Product complements create the associations that supply the companion products for any particular good that help to make its meaning clear.[30]

Mary Douglas and Baron Isherwood make a similar point: that 'Goods assembled together in ownership make physical, visible statements about the hierarchy of values to which their chooser subscribes . . . *goods in their assemblage* present a set of meanings, more or less coherent, more or less intentional. They are read by those who know the code and scan them for information.'[31] The italicized part of this quotation indicates that in their role as communicative devices, goods need to be related to one another in particular ways that on occasion may in fact allow very little room for variation (as is perhaps clearest in the case of someone wishing to convey a sense of affluence, who, for the sake of consistency, is required to purchase both a Rolex watch and a BMW car).[32]

Particularly when it is interpreted in a dynamic context of rapid product change, this form of complementarity in consumption has the potential to generate a considerable degree of frustration and disappointment. For:

> It insists that there is no such thing as a sufficiency of goods, a complement of possessions which once obtained can be considered complete. Sometimes it is dressing gowns, sometimes a new car, sometimes a new pipe, but continually we admit objects into our lives that will radically change the whole of our product complements and force us onto new levels of expenditure'.[33]

Disappointment through discovery: Hirschman's analysis of preference change

At the very beginning of Chapter 1 of his Janeway lectures, Hirschman states that his general concern is to account for preference change and he rightly emphasizes the neglect by economists of this important phenomenon. (It is somewhat difficult, though, to agree with his contention that preference change is most pronounced in Western societies, for it seems to us that a good case can be made to support precisely the opposite view that preferences are changing with particular rapidity in today's developing countries.)[34] Hirschman's more specific concern (which, as suggested below, also seems especially relevant to many parts of the Third World) is with the connection between disappointment and preference change and for this purpose he relaxes another crucial assumption of traditional theory, namely that preferences are fully known to the consumer. Instead, he envisages a world '*in which men think they want one thing and then upon getting it, find out to their dismay that they don't want it nearly as much as they thought or don't want it at all and that something else, of which they were hardly aware, is what they really want.*'[35]

Hirschman's description of this disappointment-ridden process takes as its point of departure Scitovsky's distinction between 'comfort' (the state of optimal arousal that consumption produces) and 'pleasure' (the process by which the optimal level is attained). Commodities are then classified according to the intensity with which they combine these two characteristics. At the one extreme are durable goods such as refrigerators and air-conditioners which are so highly comfort-intensive that 'pleasure is experienced only once, when the good is first acquired and put to use ("turned on"). Thereafter, comfort is assured, but the pleasure that comes with traveling from discomfort to comfort is no longer available.'[36]

At the other extreme are the 'truly non-durable goods' such as food and fuel, whose ability to provide pleasure is contingent upon their disappearance in the act of consumption itself. Hirschman contends that in the transition from the latter product category to the former (through which, we may note, many developing countries are currently passing) consumer disappointment is very likely to be experienced. The reason is that this transition involves a 'drastic' and 'unforeseen'[37] change in the balance of comfort and pleasure for the consumer who has been accustomed for a long period of time to the pleasure-intensive categories of subsistence living. That is, 'the modern, mass-produced durable consumer good subverts the way in which consumers extracted pleasure from their purchases in the predurable era. Along with their extraordinarily useful services, these goods are likely to generate a great deal of vaguely felt disappointment, particularly at the time of their first massive diffusion in society.'[38] In contrast, new products such as pharmaceuticals and soap, which, from the earlier standpoint of imperfect information, were regarded as being highly prone to disappointment, fall into Hirschman's category of 'truly non-durable goods' and as such are regarded by him as being instead especially 'disappointment-resistant'.

Although this analysis seems to bear a close resemblance to the positional process that Hirsch described, there is, as Hirschman himself points out, a rather fundamental distinction between them. For, whereas the consumer of positional goods (or, in our more specific terms, positional characteristics) '*fails* to get what he thought would be within his reach',[39] disappointments in Hirschman's terms 'are those that come from experience with goods that *are* available and affordable, but simply do not yield the sort of satisfactions which were expected from them.'[40] This distinction might, as Hirschman also suggests, bear on the way in which consumers respond on the 'supply-side' (as workers and entrepreneurs) to each of the two forms of disappointment. In particular, he suggests that failing to get what one wants (the Hirsch case) is less discouraging of future work endeavours than finding out that what one wants (and indeed gets) is not ultimately as satisfying as was anticipated. This distinction is not entirely convincing, however, for the consumer who is subject to successive bouts of (Hirschian) disappointment may in time 'get wise' to the underlying model and attenuate his productive efforts accordingly. Differential incentive effects of disappointment would seem to emerge more clearly between the form of disappointment that Hirschman emphasizes and that which in the previous section was associated with an inability of the consumer to indulge his new preferences. For what is frustrating in the latter case arises solely from inadequate income (in relation to the price of new products), which, to an arguably greater extent than is true of the disappointment associated with positional consumption, is in the hands of the consumer himself.

ACTUAL VERSUS POTENTIAL CHARACTERISTICS OF NEW PRODUCTS

Even without invoking any of the problems that have already been described in this chapter, there is still significant scope for disappointment with new products. For whereas traditional consumption theory assumes, implicitly, that consumption occurs at the point of purchase, in reality some of the most disappointing experiences with new products arise in connection with the way they are used once purchased.

More specifically, it is useful to distinguish between the characteristics that are *potentially* available from new products (which the consumer may often tend to expect) and those that are actually realized. Following Becker the household can be regarded from this point of view as a unit which produces characteristics from purchased inputs, capital and labour-time.[41] Just as in the case of production by the firm, it is possible, when dealing with the household as producer, to conceive of a maximum amount of output that is technically available from a given set of inputs. And just as in the case of the firm, there are a multitude of reasons why this maximum is unlikely to be attained in practice.

Just how large this departure from 'best practice' (within the household) can be is especially clearly illustrated in the case of modern drugs. For on the one hand, in comparison with many older drugs, which 'were by and large relatively ineffective but also relatively safe, the new drugs were often far more powerful – and far more toxic.'[42] And on the other hand, the extent to which it is the therapeutic rather than the toxic characteristics that are actually realized depends a great deal on just how the products are used. Much depends, for example, on whether the drug is taken in the correct dosage (which depends, among other things, on the age and weight of the patient), on whether it is taken at the correct time and for an appropriate length of time. Much also depends on what other drugs the patient may be taking and on the nature of the interactions between them. In each of these respects, wide variations have been observed in the behaviour of patients, not only with respect to prescription drugs, but also in the case of many over-the-counter remedies. Correspondingly wide variations have occurred in the effects of given drugs on different patients. One study, for example, revealed that while some patients were deriving more or less the expected therapeutic benefits from prescription drugs, between '4 to 35 percent of the patients were misusing their medications in such a manner as to pose serious threats to health'.[43] Even in the case of over-the-counter drugs, it has been observed that 'many patients do not follow directions, and some products . . . have caused injury or death.'[44]

In general, one might expect a greater discrepancy between the potential and realized characteristics of new products in developing than in developed countries. This for several reasons. One is related to the finding from Chapter 3 that in the vast majority of cases, multinationals do not adapt the usage instructions on the products that they sell to developing country conditions. As was also argued in that chapter, without such adaptations usage instructions are likely to be quite unsuitable for the needs of uneducated rural consumers. And this in turn means that in all likelihood, the new products concerned will yield fewer (and sometimes much fewer) characteristics than is potentially available from them. Another and related reason is what we referred to in Chapter 2 as the lack of complementarities of consumption in developing countries. Most notably, the failure of many mothers in the Third World to mix infant formula correctly is due partly to illiteracy (and their consequent inability to read the usage instructions) and partly to the absence of fuel (which is needed to boil the water to prevent contamination) or refrigeration (to prevent the rapid bacterial growth that occurs when unfinished formula is stored at room temperature).

The 'learning by consuming' concept is also relevant in this regard, for 'just as individuals and societies must learn to cope with new machines and new techniques of production, so they must learn to use consumption goods effectively. Certain items – especially consumer durables and housing – often are badly used because owners are ill-prepared to enjoy them.'[45]

SUMMARY AND CONCLUSIONS

While feelings of frustration and disappointment are never encountered in traditional consumption theory, it seems obvious from even the most casual observation that these feelings have, on not a few occasions, accompanied the consumption of new products. We have argued that there is no single mechanism through which this can occur; rather, a variety of different mechanisms can be identified. These alternatives differ, as this chapter was concerned to demonstrate, among other ways, in the nature of the commodities that yield disappointing outcomes and in the types of consumers to which these disappointing outcomes apply. These differences are shown in summary form in Table 7.1. Some of them, we suggested, may be especially pronounced in developing countries.

It needs to be emphasized, in conclusion, that these various sources of disappointment are by no means all mutually exclusive. On the contrary, in a number of cases, disappointment may be a combination of two or more of them. The infant formula case, for example, appears to be partly the result of factors that influence the initial choice of the product and partly also of the numerous difficulties that we have identified in the way that the product is used once purchased. Similarly, while positional consumption means that the positional characteristics of a product are ultimately not obtainable, the non-positional (or intrinsic) characteristics may to some extent, and for different reasons, also fail to be fully realized.

Notes

1. Cyert and De Groot (1975), p. 225.
2. Hirschman (1982), p. 17.
3. James (1983).
4. Hirschman (1982).
5. See, for example, Melrose (1982) and Chetley (1990).
6. Melrose (1982), p. 112, emphasis in original.
7. No known studies use experimental controls to assess the extent to which

Table 7.1 Varieties of consumer disappointment associated with new products

	Traditional assumption(s) rejected	*Type of new product*	*Consumers (most) affected*
Information imperfections	Complete information regarding product characteristics embodied in new goods	Pharmaceuticals, cigarettes	Those most exposed to sources of misinformation, or those with least access to accurate sources of information
Externalities in consumption	Individual satisfaction depends only on one's own consumption	Positional (typically luxury) goods; goods that pollute the environment (e.g. cars)	Those involved in the 'positional game' or those affected by environmental pollution
Preference change (a) The welfare economics of variable tastes	Preferences constant	Product embodying 'high-income' characteristics, produced or imported under high tariff protection and with a high degree of complementarity to other new goods	Those with relatively low incomes whose preferences have changed in favour of new products (and which cannot be 'indulged')
(b) Disappointment through discovery	Preferences constant and known to the consumer	Durable goods	Those able to afford durable goods
Actual vs. potential characteristics of new products	Consumption occurs at the point of purchase	Goods whose potential benefits require complementary resources and skills (e.g. infant formula)	Illiterate consumers lacking consumption skills and needed complementary inputs to the 'household production function'

advertising and marketing – as opposed to other variables – influence the use of infant formula in developing countries. One study, however, by Akin *et al.* (1985) was able econometrically to isolate the influence of sales promotion activities on infant feeding practices in the Philippines. They report, among other things, that 'A woman exposed to a facility distributing formula samples is found to be less likely to breast-feed initially and more likely to introduce breast milk substitutes in both the 7–9 and 10–12 month intervals. The important effect is probably the initial one, reducing the probability of ever breast-feeding' (Akin *et al.*, 1985, p. 76).

8. Shepherd (1985), p. 103. There are, however, some important exceptions. In 1990, Hong Kong banned television advertisements for cigarettes. Indonesia has banned smoking from primary and secondary schools, while Singapore has strengthened its already stringent regulations against the promotion of cigarettes. See *The International Herald Tribune*, 1–2 December 1990, p. 3.
9. *The Economist*, 18 May 1991.
10. *The Economist*, 2 September 1989.
11. 'It does seem to be the case', according to Hirschman, 'that the new products that are being poured out by industrial society and are finding acceptance because of their usefulness-at-first-sight will reveal whatever disadvantages or negative side effects they may harbor only slowly, as a result of repeated and prolonged use by masses of people' (Hirschman, 1982, p. 59).
12. Harsanyi (1954).
13. Ibid., p. 204.
14. Ibid.
15. On the cardinal utility approach and its critics, see Sen and Williams (1982) and Elster (1983).
16. Harsanyi (1954), p. 204, emphasis in original.
17. Ibid., p. 205.
18. Ibid., p. 208.
19. Ibid., p. 209, emphasis added.
20. Ibid., p. 213, emphasis added.
21. Harsanyi makes the point somewhat differently with the argument that 'people may be unable to tell whether a given change in their taste has *slightly* increased or slightly decreased their welfare, but they seldom seem to be in doubt as to whether a given change has *greatly* raised or greatly reduced their welfare' (Harsanyi, 1954, p. 206, emphasis in original).
22. Recall, in this regard, the finding (reported in the previous chapter) of a 'national character' effect on modernity, which suggests that modernity is influenced not merely by education, exposure to mass media and so on, but also by the average income level of the country.
23. See, for example, Landgren-Gudina (1984).
24. World Bank (1983), p. 415.
25. Ibid.
26. Kornai (1971).
27. Ibid., p. 310, emphasis in original.

28. Ibid., p. 294. Put another way, in the case of 'suction' it is possible for production to 'detach itself to a considerable extent from the consumer's aspirations' (Ibid., p. 295).
29. McCracken (1988).
30. Ibid., p. 121. McCracken refers to this as a 'Diderot effect', whose formal definition is 'a force that encourages the individual to maintain a cultural consistency in his/her complement of consumer goods' (p. 123).
31. Douglas and Isherwood (1979), p. 5, emphasis added.
32. McCracken (1988).
33. Ibid., p. 128.
34. Tse, Belk and Zhou (1989).
35. Hirschman (1982), p. 21, emphasis in original.
36. Ibid., p. 32.
37. Ibid., p. 33.
38. Ibid., p. 38.
39. Ibid., p. 61, emphasis in original.
40. Ibid., emphasis added.
41. Becker (1965).
42. Silverman and Lee (1976), p. 6.
43. Quoted in ibid., p. 318.
44. Ibid., p. 224.
45. Bruton (1985), p. 1101.

References

Akin, J.S., Griffin, C.C., Guilkey, D.K. and Popkin, B.M. (1985). 'Determinants of Infant Feeding: A Household Production Approach', *Economic Development and Cultural Change*, Vol. 34 (1).

Becker, G. (1965). 'A Theory of the Allocation of Time', *Economic Journal*, Vol. 75, Sept.

Bruton, H. (1985). 'The Search for a Development Economics', *World Development*, Vol. 13 (10/11), Oct./Nov.

Chetley, A. (1990). *A Healthy Business: World Health and the Pharmaceutical Industry*, Zed Books.

Cyert, R. and de Groot, M. (1975). 'Adaptive Utility', in Day, R. and Groves, T. (eds), *Adaptive Economic Models*, Academic Press.

Douglas, M. and Isherwood, B. (1979). *The World of Goods*, Allen Lane.

Elster, J. (1983). *Sour Grapes*, Cambridge University Press.

Harsanyi, J. (1954). 'Welfare Economics of Variable Tastes', *Review of Economic Studies*, Vol. 21 (3).

Hirschman, A. (1982). *Shifting Involvements*, Princeton University Press.

James, J. (1983). *Consumer Choice in the Third World*, Macmillan.

Kornai, J. (1971). *Anti-Equilibrium*, North-Holland.

Landgren-Gudina, M. (1984). 'Weaning Food and Low-Income Consumers in Ethiopia', in van Ginneken, W. and Baron, C. (eds), *Appropriate Products, Employment and Technology*, Macmillan.

McCracken, G. (1988). *Culture and Consumption*, Indiana University Press.

Melrose, D. (1982). *Bitter Pills: Medicines and the Third World*, Oxfam.
Sen, A. and Williams, B. (eds) (1982). *Utilitarianism and Beyond*, Cambridge University Press.
Shepherd, P.L. (1985). 'Transnational Corporations and the International Cigarette Industry', in Newfarmer, R.S. (ed.), *Profits, Progress and Poverty*, University of Notre Dame Press.
Silverman, M. and Lee, P. (1976). *Pills, Profits and Politics*, University of California Press.
Tse, D., Belk, R. and Zhou, N. (1989). 'Becoming A Consumer Society: A Longitudinal and Cross-Cultural Content Analysis of Print Ads from Hong Kong, the People's Republic of China and Taiwan', *Journal of Consumer Research*, Vol. 15, March.
World Bank (1983). *Kenya: Growth and Structural Change*, Washington DC.

Part Two
Consumption Theory and Policy in Developing Countries

8 The Promotion of Appropriate Products in Developing Countries

The chapters that comprise Part One above have a decidedly asymmetrical focus. For while they devote considerable attention to the extent and nature of the influence that is exerted on consumption patterns in developing countries by the promotion of products from the industrialized nations, these chapters pay little or no attention to the promotion of alternative (and what were described as) more appropriate patterns of consumption. From a policy point of view, this imbalance needs to be redressed partly because – as we seek first to demonstrate below – in many of the most successful cases of appropriate products in the private sector (which have recently emerged from the Indian experience), promotion appears to have played a crucial role. Furthermore, interpreted in a broader sense to include what are regarded as socially desirable product innovations (such as oral rehydration therapy), appropriate products are introduced by developing country governments over a wide range of areas (such as health, sanitation and nutrition) and the efficacy of these efforts also depends partly on the nature and form of the promotional activities that are undertaken. More specifically there is growing evidence that much can be learnt by governments in these endeavours from methods of marketing that are usually employed for commercial purposes. This issue is addressed below under the heading of what is known as 'social marketing'.

THE REMARKABLE RECENT INDIAN EXPERIENCE

In Chapter 2 we described how, for various reasons, the introduction of new products from advanced countries may give rise to the disappearance of products that are favoured by the relatively impoverished members of poor societies. And even though this extreme consequence of the introduction of new products is known to have occurred in only a few cases, a general tendency for new products to contain excessively high-income characteristics could nevertheless be documented. What is remarkable about the recent Indian experience is that in a number of sectors this tendency appears to have been substantially altered: that is, new products have been introduced which conform in essential respects to the preferences of the majority rather than the small minority of high-income consumers. In what one observer refers to as a 'revolution':

> Small and unknown manufacturers have come up from nowhere and undercut the market leaders in spectacular coups by attacking in ways and in places which others have not thought of, and which the new entrants have found the existing market leaders to be vulnerable in. In most of these cases, there was little initial investment in plant, machinery and buildings. However, *there are many times, high levels of marketing investments, usually on advertising and promotion, many times on distribution, and sometimes on sales force.*[1]

Consider first what is probably the best known of these cases, in which an indigenous brand of powdered detergent managed to capture no less than three-quarters of the market that had previously been held by Hindustan Lever, the Indian subsidiary of Unilever. The latter began production with bar soaps and cakes which were followed in turn by a soap powder and a synthetic detergent ('Surf'). Like all the other powdered detergents that were subsequently introduced in India (with the exception of 'Nirma' which we shall shortly discuss), the characteristics of 'Surf' are very much those of a typical developed-country product: in particular, it is heavily packaged in cardboard with sophisticated printing and its blue-coloured powder contains ingredients that permit optical whitening, 'kindness to hands' and so on (though it should also be pointed out that some ten years after 'Surf' was first introduced, Lever did make one concession to local circumstances in the form of polythene sachet bags in which a relatively low-priced small pack was made available).

The market dominance of 'Surf' did not go unchallenged by firms seeking to undercut the Lever product; on the contrary, many small-scale enterprises entered regional markets with low-priced detergent powders. However, these products were typically 'poorly presented, of low quality and generally unbranded'[2] and their apparent lack of success clearly emphasizes the fact that appropriateness is far from necessarily synonymous with a low price and the absence of inessential product characteristics.

It was not until the introduction of a locally produced detergent called 'Nirma' that the situation changed dramatically. Unlike 'Surf', 'Nirma' is produced in a chain of workshops which are granted significant excise tax concessions and in which non-unionized (and relatively low-cost) labourers mix the detergent powder without even electric power.[3] Being also more cheaply packaged and in other ways less sophisticated than the Lever product, Nirma is much less expensive.[4] On the other hand, what appears to distinguish the latter from other similar producers are 'a distinctive brand name and identity, created and supported by highly professional advertising and a memorable jingle. They also did not set up their own sales force but built a loyal network of distributors around the country who earned higher margins than on Surf and other brands. In this way Nirma appealed to a whole new population for whom Surf was too expensive.'[5] Indeed, so attractive to Indian consumers has 'Nirma' evidently been, that by 1988, with a turnover of $250 million,[6] this product enjoyed a market share of 60 per cent compared to the 7.1 per cent held by 'Surf'.[7]

A second, albeit somewhat less spectacular, example has occurred in the toothpaste market where a local firm, Balsara, has succeeded (where many well-known multinationals have failed) in mounting an effective challenge to 'Colgate', the market leader. The reason for this success, according to Rao, is that:

> Balsara went 'ethnic'. They recognised that cloves were the accepted remedy all over the sub-continent for problems to do with the gums, teeth and mouth odour. They therefore launched a toothpaste whose major claim was that it contained clove oil. They also made sure that the base used for the product was satisfactory, as was the foam characteristic. They had the courage to invest in advertising, despite their small size as a company. While others tried to copy Colgate or to introduce product concepts from their home markets in the US or elsewhere in the west, Balsara instead went 'native'. This success has been sustained over the years.[8]

Still other examples are to be found in the markets for packet foods and soft drinks. In the former, small-scale enterprises using labour-intensive methods of manufacturing and packaging appear to have succeeded where Lever, for all its world-wide experience in this type of market, has failed. The small-scale firms have apparently 'taken advantage of the new, cheap, simple, and attractively printed packaging material now available which ensure security from infestation of the contents, as well as attractiveness in display on the shop shelf.'[9] In the soft drinks market, a very small-scale manufacturer, Rasna, introduced powder concentrates some ten years ago.

> [This product] was easily reconstituted and the cost to the consumer per glass was significantly lower than that for carbonated beverages, squashes or other drinks. This made it appealing to the parents as well. It has a good shelf life, does not occupy space in storage, and there is no worry about storing and disposing of empty bottles for which a deposit has to be given. Rasna got its distribution by creating a demand through advertising. Coming as it did in the early days of TV network advertising, Rasna was able to get wide coverage and reach at fairly low cost.[10]

Apart from the further evidence that they provide in support of the relationship – hypothesized in Chapter 2 – between the appropriateness of technology and the appropriateness of consumption, these examples suggest that in circumstances where heavily promoted products *already exist*, particular attention may need to be paid to the promotional aspects of cheaper and more appropriate alternatives. These examples are suggestive, furthermore, in still other ways. In the first place they suggest that a distinction might usefully be drawn between different types of appropriate products. On the other hand, there are more or less traditional products of long standing which embody a very high proportion of low-income characteristics and which are sold to the poorest groups in rural communities. (The traditional Indian sugar, called gur, that is discussed in Chapter 11 is a good case in point.) On the other hand, the recent examples from the Indian experience that were cited earlier in this chapter are appropriate in another way, namely that they result from adaptations to local circumstances of *modern* products (such as synthetic detergents or toothpaste). And while it is clear that some of these products have achieved wide penetration of the market, it is doubtful that many of them are in fact benefiting the very poorest groups of consumers.[11]

The recent Indian experience also raises the important question of the role played by supply-side factors in the emergence (and in some cases the rise to prominence) of alternatives to the products of multinational corporations. For, unless these factors were such that they made profitable the sale of products such as 'Nirma', the latent demand for them would have simply gone unmet. In the case of the local manufacturer of 'Nirma' (and perhaps the other examples as well) this condition may have been satisfied by the availability of low-cost labour (on a piece-wage basis), the excise tax concessions (granted to workshops) and a somewhat less bureaucratic environment than that which large-scale enterprises have to confront.[12]

Above all, though, what these examples raise is a prospect that until now has to many seemed like mere wishful thinking on the part of those who advocate appropriate technologies and products. That is the prospect of an economy in which these principles are applied not merely at the micro (or project) level, but also at the (macro) level of the society as a whole.

SOCIAL MARKETING

Conceived some twenty years ago, social marketing represents essentially the attempt to apply commercial marketing methods to the promotion of social causes. It should be distinguished therefore both from the marketing of private goods and services and from the many ways in which information alone (as, for example, with health warnings on cigarettes) is conveyed to consumers without any attempt to invoke commercial marketing methods. As such, social marketing has been used by a variety of government and non-government institutions in both developed and developing countries.[13] In the latter context, it has been applied, among other areas, to the numerous consumption-oriented activities where government and non-government institutions have frequently sought to overcome traditional values and modes of behaviour.

One of the first such applications was to a nutrition education programme in Indonesia. 'Although mass media and other marketing methods had been employed for some years to deliver messages about nutrition, few previous attempts had been made to do the kind of research and other efforts that are widely employed in commercial marketing to meet consumer needs.'[14] These efforts included research that was designed to identify 'the beliefs, feelings, needs, constraints, and practices of mothers, with particular attention given to identifying those practices susceptible to change'.[15] On the basis of these and other inputs the project team was able to determine in quite some detail what could be achieved (for example, by the introduction of a home-made weaning food using locally available ingredients) and how best to achieve the objectives thus determined (for example, through the use of an imaginative action poster that required mothers to mark each time they performed a prescribed mode of behaviour).

According to Berg[16] this project led to improvements in nutritional status among mothers and children that could not be ascribed to other factors (such as education or radio ownership). Indeed, what is perhaps most significant about the outcome is that social marketing appears to have substituted for what is often (see Chapter 9) a major determinant of nutritional status among low-income households, namely the level of formal education of the mother. Thus, 'nutritional messages that are clear and well targeted can be understood by women who have not had the benefit of much formal education. The less-schooled mothers who were exposed to the program's specific messages demonstrated nutrition knowledge as good as mothers with more formal education and their children had as good a nutrition status.'[17]

Several other examples from the field of health and nutrition illustrate the methods used by social marketing to induce behavioural changes among the target groups of households in developing countries. In the Philippines, a project to introduce ORT (oral rehydration therapy) went much beyond simply transmitting the relevant information to mothers, an approach that has often been used in the past with very limited success. Instead, social marketing used focus groups to acquire a detailed understanding of how the ORT was actually being used by mothers. 'This consumer research revealed that some mothers dissolved only the sodium and kept the glucose as high-quality sugar for use in guests' coffee. Other mothers found the mixed ORT too salty and added sugar to make it more acceptable to their children, thereby imbalancing the treatment. These findings led to further consumer research and adjustments in the product and promotional messages.'[18] In another social marketing project conducted in Nigeria, the method employed (in this case to persuade villagers to acquire water filters as a means of controlling guinea-worm disease) again went much beyond the mere transmission of useful information; what it involved in addition was the use of group methods of persuasion (including the mobilization of local health workers and the local leader), which in other contexts (see Chapter 10) have often proven to be an effective means of inducing changes in individual values.[19]

In these and other examples that could be cited (not only from the field of health and nutrition but also from a large number of family planning projects[20]) the objective of social marketing was to introduce what are regarded as socially desirable product innovations or new practices to members of more or less traditional societies.[21] In a very different context, the social marketing approach has also been employed to exert countervailing influence against the *undesirable* effects of certain new products from advanced economies, which, as we saw in Part 1 above, are frequently heavily promoted in developing countries (Fox and Kotler in fact refer to this as 'countermarketing'[22]). Social marketing in this form has been employed, for example, to counter the adverse effects of promoting infant formula in developing countries (which were described in a number of previous chapters).[23] Another case where social marketing has been employed (to seemingly good effect) as a countervailing force concerned the promotion of coconut milk as a substitute for imported soft drinks in Micronesia. Spanning a period of two years and conducted by the Yap District Health Department:

> The project used a comic book, in-store posters, and some direct lobbying to encourage stores to stock and people to buy 'drinking coconuts'. The only newspapers on the island published a photograph of coconuts with a caption that indicated they were the 'real thing' and a well-known soft drink the artificial thing. Most young Yapese adults are literate and the slogans soon became popular catch phrases around the district, according to the evaluator. Also a series of cartoons, not originated by the nutrition project, appeared in the paper showing a canned soft drink representing unpopular foreign influence and a coconut character representing Yapese sentiment . . .
>
> As a result of the campaign, most stores on the four main islands began keeping cold coconuts in their refrigerators and selling them for half the price of soft drinks. Several individual stores reported average sales of 1,000 coconuts weekly per store on the main island, even though the total population is less than 4,000. Based on district tax receipts, imports of soft drinks to Yap decreased to less than half their previous level after the campaign.[24]

The use of social marketing in these cases is analogous to the efforts that are now being made by an increasing number of governments in the industrialized countries to 'countermarket' cigarettes, alcohol and nutritionally harmful foods of one kind or another.[25] Underlying these efforts appears to be the recognition that 'consumer sovereignty' is poorly served by a situation in which 'The large promotional budgets behind these [harmful] products tend to crowd out opposing views, because those who disagree are often too fragmented, too few in number, or have inadequate resources to present the counterposition.'[26] Indeed, in the absence of this recognition and the actions that are based on it, governments (especially in developing countries where the countervailing forces are typically especially weak) are effectively 'choosing the status quo system, *without taking the responsibility for doing so*'.[27] For what they would then effectively be assuming is that the status quo is desirable without any regard to the factors that actually give rise to it.

CONCLUSIONS

For the reasons that were given in Chapters 2 and 3, the benefits of new products tend to accrue overwhelmingly to those relatively few members of poor societies who can afford to buy them (though, as we further suggested in Chapters 6 and 7, some of these benefits are in fact only potential and may not always be fully realized in practice). The commodities purchased by the vast majority of consumers on the other hand tend to be bypassed by modern technological developments.

The recent Indian experience that was reported in the first part of this chapter is important because it demonstrates that a different pattern is possible, namely one in which the (main) benefits of modern products are provided to society by small-scale firms rather than multinationals, at a price that is far more widely affordable (albeit not low enough to be accessible by the poorest Indian households). Part of the explanation of what has occurred seems to lie on the supply side (with factors that govern the differential costs of production of small versus larger-scale firms) but much also appears to be due to the nature and extent of promotion of the substitutes for the products of the multinationals, whose dominance had previously gone largely unchallenged.

No less instructive are the various cases described in the second part of the chapter, where appropriate products have been successfully promoted (mostly by governments) as part of social marketing campaigns. Defined essentially as the application of commercial marketing methods to the promotion of social causes, social marketing appears to be effective in situations where (as in health and nutrition, for example) product innovations need to be introduced in more or less traditional societies, as well as in situations where 'countermarketing' is thought to be needed to offset the promotion of products that are harmful in the developing country context (such as infant formula).

In brief, then, what we conclude from these various findings is that a tendency to associate marketing and advertising only with inappropriate products and techniques (that can be discerned in much of the literature on these topics[28]) may be misguided. An alternative approach ought explicitly to consider the numerous ways in which these same commercial practices could be marshalled in the interests of appropriate products and techniques.

Notes

1. Rao (1989), p. M-127, emphasis added. Though our emphasis here is on small-scale firms, it is worth noting that other examples of appropriate products have emerged recently in India that are manufactured instead by large-scale enterprises. For example, collaboration between the Indian government and the Suzuki Motor Company led to the introduction of a small 800 cc car that was 'a runaway success'. (See 'Multinational Business', *The Economist*, Summer 1989). An even smaller car, with an engine size of 650 cc, is said to have been designed for the 1990s by a Brazilian entrepreneur who has already had considerable success with 'rugged jeeps and utility vehicles' in the Brazilian market. The jeeps combine fibre glass bodies with Volkswagen engines and chassis and they possess their 'own suspension and traction systems for off-the-road conditions. Therefore they are the same as a four-wheel drive at a fraction of the price' (see Kinsey, 1988, pp. 279–80).
2. Rao (1989), p. M-128.
3. *The Economist*, 14 September 1988.
4. According to Rao (1989) 'Nirma' is more 'harsh on hands' than 'Surf'.
5. Ibid., p. M-128.
6. *The Economist*, 14 September 1988.
7. Rao (1989). There is now also a 'Nirma' detergent cake which has 34 per cent of the market.
8. Ibid., p. M-128. Balsara are apparently trying to repeat this success with a new toothpaste 'Babool', which again claims to use an indigenous mouth cleanser and which is much cheaper than 'Colgate'.
9. Ibid., p. M-129.
10. Ibid., p. M-130.
11. This distinction is probably also reflected in the differential extent to which the two types of products require a redistribution of income in favour of the poor, before a significant demand for them can be created.

The traditional variety in particular is much more likely to require this form of policy than the more modern products which we have been discussing in this chapter (and which indeed have already found a substantial demand in India).

Morawetz (1974), p. 514 contends that the distinction we have just drawn corresponds to the views of different writers on appropriate products, some of whom he suggests 'are thinking of goods for which there would be high demand even with the current income-distribution pattern' while 'others implicitly assume that income will be redistributed'.

12. In some respects (and especially an emphasis on price as a competitive weapon) the Indian cases resemble certain multinationals from the Third World. In a number of cases described by Wells (1983), the latter have also been successful in capturing markets from developed-country multinationals. See also Lecraw (1981).
13. See Fox and Kotler (1980).
14. Berg (1987), p. 30.
15. Ibid.
16. Ibid.
17. Ibid., p. 32.
18. Austin (1990), p. 309.
19. Ibid. The tendency towards emulative consumption behaviour (that was described in Chapter 6) has been used on occasion with some success in the promotion of protein-enriched food. See Gordon (1973).
20. Austin (1990).
21. Under the heading of what are sometimes described as 'merit wants'.
22. Fox and Kotler (1980).
23. Ibid.
24. Hornik (1988), pp. 140–1. This reference also contains much that is useful on the subject of communications in developing countries, particularly with regard to improving nutrition.

    The example in the text is not the only occasion on which nationalistic appeals have been made on behalf of locally produced goods. Though they mostly do not constitute fully-blown social marketing campaigns, these appeals have on occasion achieved some degree of success. The introduction of a local brand of beer in Tanzania in 1976 is a case in point. What seems to have had a lot to do with the initial acceptance of this brand was a vigorous promotional campaign based on nationalism (the brand was, for example, introduced on an important national occasion and in a suitably ceremonial manner). See James (1983).
25. Fox and Kotler (1980).
26. Ibid., p. 27.
27. Barry (1965) in Penz (1986), p. 117, emphasis added.
28. In the context of the appropriate technology movement, for example, Carr (1985), p. 370, refers to a bias on the part of institutions involved in this area 'against commercial routes to the dissemination of technology . . . and in favour of the more welfare-orientated route'. She rightly suggests that while this bias is understandable in terms of the ideological foundations of the movement, it needs nevertheless to be redressed 'if a technology such as an improved stove or a more efficient brick-moulding

press is to reach the masses of poor people' (p. 371). In their book on appropriate products, van Ginneken and Baron (1984), also draw attention to the neglect of research on the marketing techniques of informal and small-scale enterprises. They suggest that more emphasis on this topic might be able to determine how these enterprises could expand their markets.

References

Austin, J. (1990). *Managing in Developing Countries*, Free Press.

Barry, B. (1965). *Political Argument*, Routledge & Kegan Paul.

Berg, A. (1987). *Malnutrition: What Can be Done*, Johns Hopkins University Press.

Carr, M. (1985). *The AT Reader*, Intermediate Technology Publications.

Fox, K. and Kotler, P. (1980). 'The Marketing of Social Causes', *Journal of Marketing*, Vol. 44, Fall.

Ginneken, W. van and Baron, C. (eds) (1984). *Appropriate Products, Employment and Technology*, Macmillan.

Gordon, W. (1973). 'Institutionalized Consumption Patterns in Underdeveloped Countries', *Journal of Economic Issues*, Vol. VII, No. 2, June.

Hornik, R. (1988). *Development Communication*, Longman.

James, J. (1983). *Bureaucratic, Engineering and Economic Men: Decision-Making for Technology in Tanzania's State-Owned Enterprises*, ILO, World Employment Programme Research, Working Paper No. 125.

Kinsey, J, (1988). *Marketing in Developing Countries*, Macmillan.

Lecraw, D. (1981). 'Internationalization of Firms from LDCs: Evidence from the ASEAN Region', in Kumar, K. (ed.), *Multinationals from Developing Countries*, Lexington Books.

Morawetz, D. (1974). 'Employment Implications of Industrialization: A Survey', *Economic Journal*, Vol. 84 (335), Sept.

Penz, P. (1986). *Consumer Sovereignty and Human Interests*, Cambridge University Press.

Rao, S.L. (1989). 'Innovative Marketing Strategies: Small Enterprises Fight Large Established Companies', *Economic and Political Weekly*, 26 August.

Wells, L. (1983). *Third World Multinationals*, MIT Press.

9 The New Household Economics, General X-Efficiency Theory and Consumption in Developing Countries

Associated mainly with the work of Gary Becker, the 'new household economics' views consumption as a process of 'producing' characteristics from purchased inputs and labour time.[1] As such, it establishes a pioneering link between consumption and the theory of production. Among other things, this analogy enables one to analyse the efficiency with which the various activities in the household, such as cleaning, administering health care, cooking and feeding, are carried out. For, just as production by the firm may be more or less efficient, so too can households vary in the efficiency with which they conduct their various activities. When they are applied to activities that are closely related to basic needs (such as infant feeding or the provision of health care), these variations may help to explain the differential extent to which poor households are able to meet the basic needs (for hunger, thirst and so on) of their members and, more specifically, why education seems to be so important in accounting for these differences.

At the same time, however, it is essential to recognize that in addressing important policy-related questions such as these, the new household economics suffers from some serious limitations. In particular, the household as producer is assumed to behave according to the simplest tenets of neoclassical micro-theory, some of which (such as the assumption of given and constant preferences) we have already had much cause to criticize in earlier chapters. The purpose of this chapter is to suggest that an alternative approach, based on X-efficiency theory, provides a broader and behaviourally more realistic framework for analysing household activities in general and those that are closely related to basic needs in particular.

THE NEW HOUSEHOLD ECONOMICS VERSUS GENERAL X-EFFICIENCY THEORY

Briefly stated, Becker's approach to consumption:

> . . . transforms the family from a passive maximizer of the utility from market purchases into an active maximizer also engaged in extensive production and investment activities. In the traditional theory, households maximize a utility function of the goods and services bought in the marketplace, whereas in the reformulation they maximize a utility function of objects of choice, called commodities, that they produce with market goods, their own time, their skills, training and other human capital and other inputs.[2]

What is innovative about this approach is its transformed view of the household (as producer rather than consumer); in other respects, however, it remains squarely in the tradition of neoclassical microeconomics and conventional welfare economics. For not only is maximization assumed to take place in the usual way, but what is to be maximized is taken to be 'a unified family welfare function' that represents the collective (and implicitly harmonious) interests of the household members. Furthermore, in the time-honoured tradition of conventional consumption theory, the new household economics assumes that preferences – for the various 'commodities' that the household produces – are given and constant.

The usefulness of the new household economics as an analytical tool and as a basis for policy formulation in the area of basic needs depends on the degree to which these assumptions capture the realities of household behaviour among poor people in poor countries. Our view is that they do so to only a very limited degree and that the behavioural assumptions posited by general X-efficiency theory provide in many ways a more realistic alternative. It is worth emphasizing, though, that the proposed framework does not imply a rejection of *all* the elements of the new household model. Rather, what is suggested is an integration of X-efficiency theory with the fundamental insight (provided by Becker) of the household as a producing unit.

General X-efficiency theory differs from conventional microeconomics in the following major respects:

1. its psychological assumptions;
2. it allows for the possibility of conflicting interests within the household, including those that arise in the relationship between agents and principals;
3. it lays considerable emphasis on the theory of inert areas.

PSYCHOLOGICAL ASSUMPTIONS

The point of departure for Leibenstein's X-efficiency theory[3] is the view that existing micro-theory relies on an assumption about individual behaviour that is psychologically untenable. Perhaps because it seems self-evident, maximization is often merely *assumed* in traditional theory. But, according to Leibenstein, once the behavioural assumptions underlying the maximization postulate are scrutinized, they turn out to make excessive psychological demands on the 'typical' individual. In particular, he regards as entirely unrealistic the view of decision-making in traditional micro-theory as effortless and costless.

There are a number of dimensions along which individuals operate in pursuance of maximizing behaviour. One of them is 'calculatedness', which can be viewed along a continuum ranging from 'loose' or 'sloppy' calculation at one end, to 'tight' calculation at the other. At the former end 'the degree of calculation is so sloppy as to be almost the same as a random choice.'[4] At the latter end, 'it is very careful – continually checked and rechecked, so that only a completely accurate answer, if possible, is tenaciously sought.'[5] Some individuals (such as 'perfectionists') will always value the tight over the sloppy calculation and there will also be some circumstances in which most individuals prefer the tight calculation, but it seems plausible to argue that 'for some people at least some of the time, the comfortable degree of calculatedness is not very tight.'[6]

On other dimensions, too, individuals will find the 'tight' procedures required for maximization to be psychologically uncomfortable. It is rarely true, for example, that people have the high degree of sensitivity to changes in magnitude implied by maximization. Rather: 'Magnitude sensitivity involves taste to some degree. Psychologically we may be either insensitive to magnitudes within certain bounds or we may find it unpleasant to be so.'[7]

In general, according to X-efficiency theory, individuals in choosing a comfortable rather than a highly tight position along the various

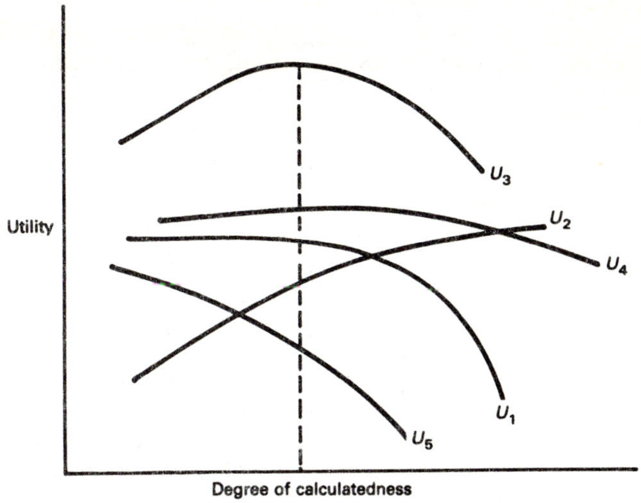

Figure 9.1 The equilibrium degree of calculatedness.
Note: U_4 and U_5 are alternatives to U_1.

dimensions of decision behaviour behave according to a selective rationality rather than according to the 'full rationality' underlying the behaviour of 'economic man'.

The degree of selective rationality chosen depends upon the extent of disutility associated with various levels of 'calculatedness' in relation to the perceived incremental benefits of making a choice based on 'tight' calculations. Diagramatically as shown in Figure 9.1 the matter may be viewed as follows. The degree of calculatedness is assumed to be meaningful (and measureable) in an ordinal sense only, and the curve U_1 measures the utility associated with varying degrees of calculatedness. The non-linearity of this downward sloping curve reflects the fact that beyond a certain point the disutility associated with higher degrees of calculatedness increases quite sharply. Curve U_2 measures the utility obtained from the additional characteristics that are obtainable from choices that are closer to full rationality, that is from greater calculatedness. Following the usual assumption of declining marginal utility as consumption increases, the curve increases at a decreasing rate. The addition of curves U_1 and U_2 yields curve U_3, with the maximum of which the equilibrium degree of calculatedness is associated.

The degree of calculatedness that is chosen by the consumer will be

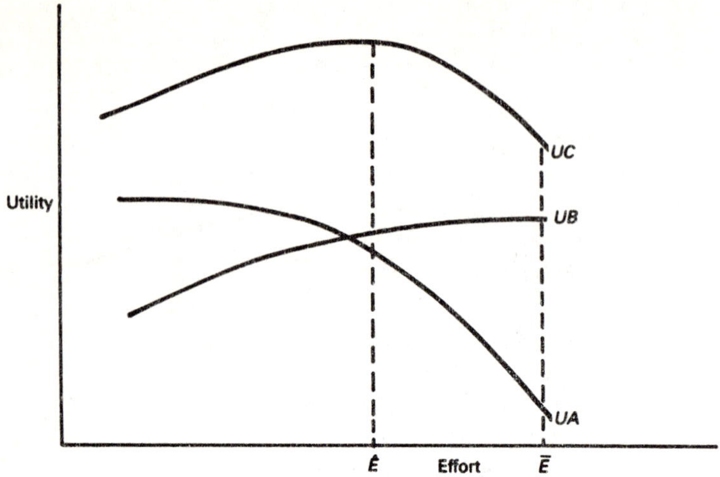

Figure 9.2 The equilibrium degree of effort.

one of the factors that determine the efficiency with which goods (inputs in the Becker model) are purchased by the household. However, following X-efficiency theory one also needs to investigate the extent to which the chosen inputs are efficiently *used* by the household. The nutrients obtained from food, for example, depend heavily on how it is actually prepared; one study found nutrient losses of as much as 40 per cent due to the way traditional food was prepared in Nigeria.[8] And this in turn depends, so the theory goes, to a considerable extent, on the degree of (mental and physical) effort that is exerted in the process. Figure 9.2 conceptualizes this issue in a manner that is similar to Figure 9.1.

The horizontal axis measures (ordinally) the amount of effort (both mental and physical) in converting inputs to characteristics. The relationship between utility and effort shown in curve *UA* is entirely analogous to curve U_1 in Figure 9.1. In addition, curve *UB* – which shows the utility derived from the extra characteristics obtained as a result of increased effort – is similar to curve U_2. Here as in Figure 9.1, of course, personality and societal factors influence the equilibrium position. Malnourished (and often diseased) individuals in developing countries, for instance, are likely to find effort, beyond some point, considerably more onerous than the majority of those living in more affluent societies who are in good mental and physical health.

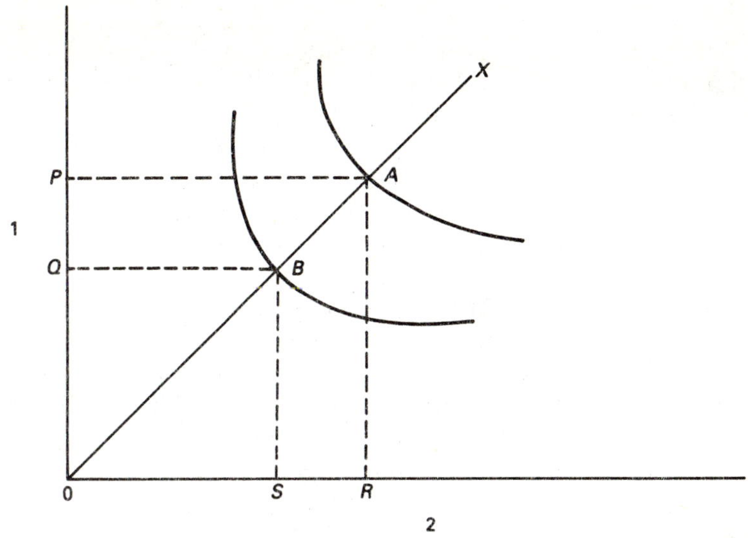

Figure 9.3 A measure of X-inefficiency.

The former, *ceteris paribus*, will therefore tend to display a lower degree of effort in equilibrium than the latter.

Whether X-inefficiency occurs and the extent to which it occurs depends upon the slopes of the curves *UA* and *UB* (and hence the equilibrium level of effort) in relation to the *required effort level*. Some goods require much more effort than others to realize all their available characteristics. Goods differ, for instance, in the attention to detail and the coordination of tasks that they require for efficient conversion into characteristics. Where the required effort level (say $\bar{E}$) is high in relation to the equilibrium level ($\hat{E}$), a substantial degree of X-efficiency may be expected. This magnitude can be illustrated with reference to Figure 9.3.

Assume, for the sake of argument, that good *X* is capable of yielding (for a given expenditure) 0*P* units of characteristic 1 and 0*R* units of characteristic 2, if it is used in the way that its manufacturer intended. One could, for example, think of this product as a medical drug , whose efficacy (measured therapeutically by the characteristics 1 and 2) demands that usage instructions be closely followed. With less than the degree of effort that this requires, however, the consumer may end up with far fewer of the potential benefits (say, the amounts 0*Q* of characteristic 1 and 0*S* of characteristic 2). The difference between the potential and the actual amounts of characteristics is then a measure of X-inefficiency in consumption.[8] The examples that were provided in Chapter 7 illustrate all too clearly how large this differential can actually be.

Unlike the firm, which has been intensively studied, especially in the context of developed countries, X-inefficiency within the household has received very little attention (either in developed or developing countries). Yet, there is no a priori reason to expect the extent of the losses from this cause to be any less serious in the household than in the firm (where, according to some estimates, the overall level of X-inefficiency may be as high as 40 per cent of net national product). And there is reason to think that the problem might be especially pronounced in poor, uneducated, rural households in developing countries. For, not only may motivation in these households be impaired by nutritional and health deficiencies (as noted above), but it may also be adversely affected by a range of sociological and psychological variables. There is a large body of literature, for example, which has studied the way in which traditional values influence a number of motivational factors (such as achievement motivation,[9] openness to new ideas and one's sense of efficacy). We shall refer to some of this literature below in the section dealing with the role of inert areas in X-efficiency theory.

INTRA-HOUSEHOLD CONFLICTS OF INTEREST

The new household economics entirely skirts the questions of how conflicts may arise within the household and how these conflicts ultimately get resolved. These questions are able to be avoided by the use of concepts such as a 'well-behaved nuclear family utility function'[10] (which assumes that 'the family can be approximated as a unified optimizing consumer'[11] and a 'benevolent dictator'[12] (whose utility function is presumed to reflect the interests of the rest of the household).

X-efficiency theory, in contrast, allows for the possibility of conflict within the productive unit, whether it be a firm or a household. The theory pays particular attention to the distinction between agents and principals: 'Agents may make decisions in the interests of their principals, but they make decisions which emphasize their own interests irrespective of whether they are in the principals' interest. There is no reason to presume that the contractual relationship between agents and principals is either a complete one or is one that necessarily prevents the agents from deviating from maximizing the principals' interests.'[13] In fact, X-efficiency theory posits that in general the individual will be more selectively rational – with a greater deviation from the effort associated with maximization – if he is pursuing someone else's rather than his own interests.

So far, this distinction has mostly been applied to the firm and, more specifically, to explain what so often seems to be the considerable degree of X-efficiency within it. As in the case of the firm, however, relationships between members of the household may also be important in determining the extent of X-inefficiency. In many cases, for example, the housewife is acting as an agent in that she is purchasing goods and converting them into 'commodities' whose enjoyment will be experienced by others as well as herself. Although we may perhaps expect a greater coincidence of interests between agents and principals within the household than within the somewhat more impersonal context of the firm, there is no reason why there should always be a complete identity of interests. Indeed, a considerable sociological literature attests to the often conflictual nature of intra-household relationships.[14] And the way in which these conflicts are resolved may often have a crucial bearing on the well-being of different family members.

For instance, in a study of family influences on eating habits, Graham has observed that despite the strenuous efforts made to persuade mothers to delay the use of solids and adopt alternative solutions to their feeding problems which were less detrimental to their babies' health, these solutions were not adopted. What she concluded was that the alternative solutions, 'placed too high a premium on the baby's health and in consequence, too high a cost on the mother and other members of the family. Like breastfeeding, giving water was a baby-oriented method of feeding, which, in the context of the mother's conflicting family commitments, was unworkable.'[15] That is, infant feeding has to be seen in the wider context of intra-household pressures and conflicts, and inefficiency therein cannot be understood by focusing on the mother herself.

More generally, the resolution of household conflicts often appears to disadvantage women who 'subordinate their own needs and choices in fulfillment of what is expected of them as a mother and wife'.[16] On the other hand, 'Men . . . are often under less social and cultural pressure to subordinate their basic needs to those of other family members – except perhaps in situations of severe poverty or household crisis. Male roles tend to be more independently constituted than female roles, and men are often able to make choices that lie outside any obligations to family and household welfare.'[17]

INERT AREAS AND THE RESISTANCE TO CHANGE OF TRADITIONAL VALUES

In contrast to traditional microeconomic theory, which has no such concept, inert areas play a crucial dynamic role in X-efficiency theory because it is concerned with the extent to which opportunities for change from a *given* chosen position are actually pursued. Because of inertial cost, namely the psychological cost of moving *away* from the given position as well as of 'settling into' the new position, not all opportunities for change will be exploited. In particular, those opportunities which do not result in a gain of utility (or avoidance of a loss) in excess of the cost of the move will not take place. The importance of inert area theory derives from the fact that it 'allows for a degree of stability in individual behaviour in the face of a great many changing external conditions.'[18] Not only, therefore, may the individual choose a position that corresponds to selective rather than complete rationality, but he may also fail to deviate from the non-maximizing position in spite of apparently favourable opportunities for change.

In the context of basic needs in developing countries, and from the standpoint of the policy-maker seeking to meet these needs, one particular application of inert area theory assumes great importance, namely the powerful inertial forces that characterize and indeed maintain traditional values and behaviour among the poor (though this importance is by no means always recognized in policy-making, which, partly as a result, frequently fails to meet its objective of poverty alleviation). Galbraith has perceptively described these inertial forces in terms of what he refers to as the 'accomodation' by the poor to their poverty.[19] Noting how the 'will to improve' one's lot is taken for granted in much of economic policy, Galbraith argues not only that this type of motivation is generally absent from poor communities of long standing, but also that its absence has a profoundly rational explanation. Thus:

> People who have lived for centuries in poverty in the relative isolation of the rural village have come to terms with this existence. It would be astonishing were it otherwise. People do not strive,

> generation after generation, century after century, against circumstances that are so constituted as to defeat them. They accept. Nor is such acceptance a sign of weakness of character. Rather, it is a profoundly rational response. Given the formidable hold of the equilibrium of poverty within which they live, accomodation is the optimal solution. Poverty is cruel. A continuing struggle to escape that is continuously frustrated is more cruel. It is more civilized, more intelligent, as well as more plausible, that people, out of the experience of centuries, should reconcile themselves to what has for so long been the inevitable.[20]

Traditional consumption theory in general, and the new household economics in particular, afford little assistance to the policy-maker who is concerned to understand and to overcome the powerful inertial forces that Galbraith describes (in areas such as, for example, health, nutrition and sanitation). For, as Hirschman has rightly emphasized, to the limited extent that this theory does concern itself with preference change, it is only with the relatively minor, or 'wanton' variety.[21] 'These are the preference changes economists have primarily focused on: impulsive, uncomplicated, haphazard, publicity-induced, and generally minor (apples vs. pears) changes in tastes.'[22] What is neglected, according to Hirschman, is preference change of a more fundamental kind: this, he refers to as 'the non-wanton change of preference', which, in fact, 'is not really a change in tastes at all'.[23] For, 'A taste is almost defined as a preference about which you do not argue – *de gustibus non est disputandum*. A taste about which you argue, with others or *yourself*, ceases ipso facto being a taste – it turns into *a value*.'[24] The neglect by economists (among whom advocates of the new household economics such as Becker and Stigler are singled out for particular criticism) of this second type of preference change has, according to Hirschman, had the unfortunate consequence of impeding 'a serious intellectual effort to understand what are strongly held values and difficult to achieve changes in values rather than tastes and changes in tastes.'[25]

It seems clear to us that in any such effort (which acquires a particular urgency in the developing country context where governments, partly under the heading of 'merit wants', seek to alter household behaviour across a wide range of activity) social psychology will play an important part. Indeed, although he does not allude to it, it is possible to interpret much of Galbraith's accommodation to poverty in terms of the psychological theory of cognitive dissonance that we have referred to on several occasions in the previous chapters (and which is discussed in detail in the following chapter). In these terms, the individual would find it all but impossible to reconcile the conflict between his desire to escape from the grinding poverty in which he finds himself with the recognition of the futility of any such efforts. As a result, so the theory goes, he would tend to seek one way or another of lessening or eliminating the conflict (that is, of rationalizing the immutability of his plight). This could take the form, for example, of diminishing the desirability of what is unattainable (the so-called 'sour grapes' phenomenon[26]), or of taking a more favourable view of what *is* attainable (or, as it is sometimes expressed, 'being grateful for small mercies').

It seems to us no less clear that in any study of the resistance to change of traditional values, sociology and politics have a major contribution to make. For unlike tastes (in Hirschman's sense), which may often change whimsically without much regard for the behaviour of others, changes in an individual's values are usually strongly conditioned by the norms of the wider society to which he belongs. Some of the most important of these mechanisms (which involve both sociological and political factors) have been usefully summarized by Robin Williams in relation to 'values especially resistant to change at the individual level'.[27] These he suggests:

> . . . seem to involve psychologically '*primitive*' beliefs for a high proportion of social actors, and are highly *interconnected* with other values for a high proportion of the individuals who endorse them. In the social system, change-resistant values are high in *centrality*, *pervasive*, and supported by powerful *sanctions* and high *consensus*; supporters of these values hold positions of high *prestige* and authority. In the cultural system, the most resistant values are *congruent* with many other values and beliefs, and are *symbolized* in many different ways. Values having these 10 characteristics will be strongly sanctioned and actively defended and promoted.[28]

Consumption-oriented policies that do not address these various sources of resistance to change are not likely to be very successful, as the failed attempt to introduce water-boiling in a Peruvian village illustrates.[29] Despite an intensive two-year campaign to persuade villagers to boil drinking water (as an important health practice), the rate of adoption of this new practice was only about six per cent. Part of the reason for this disappointing outcome can be ascribed to the:

> ... cultural beliefs of the villagers. Local tradition links hot food with illness. Boiling water makes it less 'cold', and hence, appropriate only for the sick. But if a person is not ill, he is prohibited by village norms from drinking boiled water. Only the individuals who are least integrated into local networks risk defying community norms on water boiling. An important factor affecting the adoption rate of any innovation is its compatibility with the values, beliefs, and past experiences of the social system.[30]

Much can also be learnt from the (mainly sociological) literature on modernization that was discussed in Chapter 6. Not only does this literature (as shown there) offer a systematic analysis of the way in which traditional values change over time, but it also applies the results of this analysis to various policy concerns. One of them is the acceptance of family planning, which, perhaps more than any other area of consumption-oriented policy, has to confront the resistance of traditional values. The study conducted by Inkeles and his associates shows, as one might expect, that the acceptance of family planning is closely related to what in Chapter 6 was referred to as the 'modernity syndrome' (higher living standards, in contrast, appear to have very little influence on this decision). However, not all the components of the 'syndrome' that were listed in that chapter have an equal influence (in the regression results) on the acceptance of family planning. Rather, in each of the four diverse cultures (East Pakistan, India, Israel and Nigeria) from which the sample was drawn, the 'valuation of science, technology and medicine is the component of attitudinal modernity which has the greatest impact'.[31] Inkeles concludes that this variable should be accorded particular attention in policies designed to promote the acceptance of family planning. 'For example, through films, film strips, and illustrated literature, the point should be emphasized that man in general, and each individual in particular, can solve the problem of unwanted births just as man has conquered many diseases, increased food production, and invented machines which can do the work of many men.'[32]

EDUCATION, THE HOUSEHOLD AND NUTRITION

Education, as noted earlier, has a strong positive correlation with basic needs indicators, such as life expectancy.[33] However, the nature of this relationship – how it is transmitted through household activities – is still far from clear. In what follows, we try to show how elements of the framework described above can help to clarify this question, in the specific context of food and nutrition. The choice of this context reflects in part the fact that unlike some other products that are related to basic needs, some of the most important characteristics of food (namely, its nutritional components) are relatively easily measured. And in relation to these characteristics at least, the focus of much of this chapter on questions of household efficiency can be illustrated and justified.

The evidence on which we shall rely is drawn from a detailed econometric analysis by Berhman and Wolfe,[34] which examines the determinants of one aspect of consumption efficiency, namely the efficiency with which food inputs are purchased by the household. Behrman and Wolfe show first that the income elasticities for nutrient intakes are relatively low and they suggest that it is accordingly 'misleading to assert that generally in developing countries income improvements will lead to substantial improvements in nutrient intakes, as the World Bank . . . and others have claimed.'[35] Women's schooling in contrast is shown to have a favourable impact on nutrition that is widespread and substantial. Berhman and Wolfe doubt, however, that this favourable impact was transmitted solely through the mechanism that is posited by the new household economics, namely through increased productivity in household activities. They conclude that other ('non-efficiency') factors are probably also of importance. In particular, 'women with more schooling have greater *concern* about nutrition or are better able to implement their concerns, so their households tend to have better nutrient inputs, *ceteris paribus*.'[36]

The importance of these two explanatory variables (both of which fall outside the purview of the new household economics) at once illustrates and confirms the central theme of this chapter, namely that a far broader approach to the household is required than that which is offered by the new household economics. For both the increased concern with nutrition and the greater degree of power within the household of more educated women, can be interpreted in terms of the alternative framework (based on X-efficiency theory) that this chapter has advocated. The former bears on the distinction that is ignored in the new household economics, namely the difference emphasized by Hirschman between changes in tastes and changes in values. In particular, the increased concern with nutrition shown by the more educated women can reasonably be regarded as part of the changes in values that education is known to produce (see Chapter 6). On the other hand, the increased power of the more educated women within the household attests to the importance of intra-household conflicts and the manner in which they are resolved (neither of which, it will be recalled, is addressed by the new household economics).

It needs to be emphasized though that the results reported by Behrman and Wolfe deal with only one link in the complex relationship between food (as inputs to the household production process) and the satisfaction of basic needs: namely the link between food purchases and the nutrients embodied in these purchases. What is not dealt with by them is the efficiency of the process by which food purchases are converted (for example, by storage and cooking) into *actual* nutrients. This relationship is crucial because it appears to be the case (as noted above) that a *given* set of inputs (including food) can yield very different nutritional outcomes (depending, for example, on how the food is prepared). But even this relationship does not go far enough: a third link needs still to be considered in determining the ultimate impact of food on the basic needs of those living in poverty. This link concerns the effect of the 'realized' nutritional characteristics (in the sense just described) on individuals whose personal circumstances may vary greatly from one case to another. As Sen has put it, 'The characteristics of the goods do not tell us what the person will be able to do with those properties. For example, if a person has a parasitic disease that makes the absorption of nutrients difficult, then that person may suffer from undernourishment even though he may consume the same amount of food as another person for whom that food is more than adequate.'[37] Sen rightly argues that any judgment of the well-being of the person cannot be limited to the analysis of the characteristics of the goods that he or she possesses. What is required in addition is consideration of the 'functionings' of persons (which, in brief, refer to what the individual is able to do with the commodities and characteristics at his or her disposal). Sen's important concept of 'functionings' takes us still further from the methodological limitations imposed by the new household economics.[38]

Enlightening though it often may be, decomposition of the relationship between commodities and 'functionings' into the two categories that have just been described (namely goods to characteristics and characteristics to 'functionings'), will not, however, always be possible.

For, unlike those pertaining to nutrition, product characteristics (in the realm of basic needs as elsewhere) may on occasion be subjective or otherwise difficult to quantify. In such cases, it will accordingly be necessary to go instead straight from commodities to 'functionings'.

CONCLUSIONS

This chapter has argued that, as a foundation for consumption policy in areas such as health and nutrition, the new household economics suffers from a number of serious limitations. These limitations mean (among other things) that the new household model is unable to capture the manifold mechanisms through which education plays such an important role in the satisfaction of basic needs in developing countries. We suggested that an alternative approach, based on general X-efficiency theory, provides a broader and behaviourally more realistic framework in which these and other important policy issues in consumption can be addressed.

Notes

1. Becker (1965).
2. Stigler and Becker (1977), p. 77.
3. Leibenstein (1980).
4. Ibid., p. 73.
5. Ibid.
6. Ibid.
7. Ibid., p. 85.
8. Stewart (1985). Interestingly, the nutrients derived from food may also depend on the manner in which it is actually eaten. It has been argued, for example, that the (traditional Peruvian) practice of beginning meals with foods which have to be chewed and require greater salivation 'converts the food into a chemically treated pap which, upon reaching the stomach, is easily broken down by gastric juices, facilitating the assimilation of nutrients.' (See *Third World Resurgence*, No. 6, p. 10.)
9. McClelland (1961).
10. Schultz (1988), p. 428.
11. Ibid.
12. Evans (1989), p. 13.
13. Leibenstein (1978), p. 27.
14. Evans (1989) and Ferber and Birnbaum (1977).
15. Graham (1980), p. 176.
16. Evans (1989), p. 12.

17. Ibid., p. 13.
18. Leibenstein (1980), p. 114.
19. Galbraith (1979).
20. Ibid., pp. 62–3.
21. Hirschman (1986), p. 145.
22. Ibid.
23. Ibid.
24. Ibid., emphasis in original.
25. Ibid.
26. Elster (1983).
27. Williams (1979), p. 33.
28. Ibid., pp. 33–4, emphasis in original.
29. Rogers (1983).
30. Ibid., p. 4.
31. Inkeles (1983), p. 223.
32. Ibid.
33. See Stewart (1985) for a summary of these results.
34. Behrman and Wolfe (1984).
35. Ibid., p. 127.
36. Ibid., p. 118.
37. Sen (1985), p. 9.
38. See also the concept of the 'meta-production function' in Stewart (1985).

References

Becker, G. (1965). 'A Theory of the Allocation of Time', *Economic Journal*, Vol. 75, Sept.

Behrman, J. and Wolfe, B. (1984). 'More Evidence on Nutrition Demand: Income Seems Overrated and Women's Schooling Underemphasized', *Journal of Development Economics*, Vol. 14.

Elster, J. (1983). *Sour Grapes*, Cambridge University Press.

Evans, A. (1989). *Gender Issues in Rural Household Economics*, Institute of Development Studies, Discussion Paper 254, University of Sussex.

Ferber, M. and Birnbaum, B. (1977). 'The "New Home Economics": Retrospects and Prospects', *Journal of Consumer Research*, Vol. 4.

Galbraith, J.K. (1979). *The Nature of Mass Poverty*, Harvard University Press.

Graham, H. (1980). 'Family Influences on the Eating Habits of Children', in Turner, M. (ed.), *Nutrition and Lifestyles*, Applied Science Publishers.

Hirschman, A. (1986). *Rival Views of Market Society and Other Recent Essays*, Viking.

Inkeles, A. (1983). *Exploring Individual Modernity*, Columbia University Press.

Leibenstein, H. (1978). *General X-Efficiency Theory and Economic Development*, Oxford University Press.

Leibenstein, H. (1980). *Beyond Economic Man*, Harvard University Press.

McClelland, D. (1961). *The Achieving Society*, Van Nostrand.

Rogers, E. (1983). *The Diffusion of Innovations*, 3rd edn, Free Press.

Schultz, T.P. (1988). 'Economic Demography and Development: New Directions in an Old Field', in Ranis, G. and Schultz, T.P. (eds), *The State of Development Economics*, Blackwell.

Sen, A. (1985). *Commodities and Capabilities*, North Holland.

Stewart, F. (1985). *Planning to Meet Basic Needs*, Macmillan.

Stigler, G. and Becker, G. (1977). 'De Gustibus Non Est Disputandum', *American Economic Review*, March.

Williams, R. (1979). 'Change and Stability in Values and Value Systems: A Sociological Perspective', in Rokeach, M. (ed.), *Understanding Human Values*, Free Press.

10 Cognitive Dissonance Theory and Consumption Policy

The niceties of the Becker/Stigler argument that individual preferences are the same and constant will be of little interest to the policy-maker in a developing country whose task is to *alter* norms in areas such as health, nutrition, family planning and sanitation.[1] In most such areas, those who formulate policy will have to confront what are often deep-seated and unyielding values, about which, as noted in the previous chapter, consumption theory of the Becker/Stigler variety has little or nothing to say.

In this endeavour policy-makers would do better to turn to the evidence accumulated in other disciplines that do concern themselves with the question of how human values are formed and how they change over time. We have already referred in this regard to the mainly sociological literature on modernization which not only seeks to provide a systematic analysis of the ways in which traditional values change over time, but also applies these findings to important policy issues (such as for example the diffusion of methods of family planning). Social psychology also has much to contribute to these issues although, curiously, it has not been applied in the development context as widely as one might have expected.[2] Indeed, the theory of cognitive dissonance, which has had such an enormous impact on the way that social psychologists think about the relationships between motivation and behaviour, has been almost totally ignored in development studies (apart that is, from the treatment given to it by Hirschman in various contexts and at different times[3]). As we see it, this is a regrettable state of affairs, especially since the dissonance formulation has certain direct and important implications for the design of policies to promote the kinds of values that are known to be essential for the process of development. The purpose of this chapter accordingly, is to analyse some of the developmental implications of this well-established theory and, through an interpretation of a selected set of consumption-oriented policies in the light of these implications, to suggest areas for improvement in policy-making.

Following a summary of Hirschman's original analysis of the issue, we indicate the respects in which it is proposed to extend his analysis.

In the next section, the theoretical underpinnings of the chapter are presented. This is followed by a set of empirical applications of the argument, on the basis of which certain policy implications can be drawn.

HIRSCHMAN'S ANALYSIS

Hirschman[4] was the first to draw attention to the implications of cognitive dissonance theory for the process of what he terms attitude change in development, and in particular to the key implication of the theory, that a change in attitudes may *follow*, rather than *precede* behavioural change. (Hirschman's use of the term 'attitudes', in this context, seems to conform closely to the notion of values with which we are here mainly concerned.) Thus:

> The theory states that a person who, for some reason, commits himself to act in a manner contrary to his beliefs, or to what he believes to be his beliefs, is in a state of dissonance. Such a state is unpleasant, and the person will attempt to reduce dissonance. Since the discrepant behavior has already taken place and cannot be undone, while the belief can be changed, reduction of dissonance can be achieved principally by changing one's beliefs in the direction of greater harmony with the action . . .
>
> Dissonance theory deals with the possibility of replacing the 'orderly' sequence, where attitude change is conceived as the prerequisite to behavioral change, by a 'disorderly one', where modern attitudes are acquired *ex-post*, as a consequence of the dissonance aroused by 'modern' type of behavior which happens to be engaged in by people with non-modern attitudes.[5]

The problem is, of course, one of inducing precisely such behaviour on the part of people whose attitudes are fundamentally in conflict with it. For Hirschman, however:

> this is not much of a problem among *late coming* societies surrounded by modernity and by opportunities to transgress into or try out modern behavior; at one time or another, it is likely that the latecomer *will stumble more or less absent-mindedly into such behavior* as pursuit of individual profit, entrepreneurial risk-taking, promotion according to merit, long term planning, holding of

> democratic elections, etc.; dissonance will thus arise and will then gradually lead to those changes in attitude and basic beliefs which were thought to be prerequisites to the just-mentioned modes of behavior.[6]

The role of government, accordingly, is primarily to promote the extent to which individuals are able to 'engage in these dissonance-arousing actions and in inducing initial commitment to them'.

Hirschman's analysis does not, however, adequately incorporate the fact that the inducement of 'dissonance-arousing actions' by government can – and indeed often does – take a wide variety of forms, ranging from mild encouragement on the one hand, to outright coercion on the other. It is the contribution of dissonance theory to show that these alternatives, and the precise manner in which they are effected, will have very different implications for the extent to which a favourable change in attitudes can be expected to accompany induced alterations in behaviour. Our concern in this chapter, therefore, is to consider in more detail than does Hirschman, the implications of the theory for the probable efficacy of government policies that seek, directly or indirectly, to induce people to alter their consumption-oriented behaviour. For this it is necessary to consider in some detail (1) the conditions under which cognitive dissonance will be aroused by certain types of behaviour and (2) how the dissonance thus aroused will be dealt with by an accompanying change in attitudes.

THE CONDITIONS FOR ACCOMPANYING ATTITUDE CHANGE CONSEQUENT UPON BEHAVIOURAL CHANGE

'According to the dissonance theory analysis, behaving in a manner contrary to one's attitudinal position creates dissonance, which can be reduced by changing one's attitudes so that they become more consistent with the behavior engaged in.'[7] This sequence, however, is by no means automatic – on the contrary, dissonance theory lays down a quite specific set of conditions that must be fulfilled in order for it to occur. In this section we examine the nature of these conditions.

There are two distinct stages in the process by which behaviour change produces a favourable change in attitude through the dissonance mechanism. Each stage will be seen to give rise to a necessary condition for this change to occur. First, it is necessary that the behavioural change actually induces dissonance, and secondly, it is required that attitude change is the method that is chosen to reduce the dissonance. Together, the two sets of conditions will be sufficient for the required attitude change.

The conditions for dissonance arousal

The basic mechanism that induces dissonance as a consequence of behavioural change is one of conflict, and in particular the conflict between one's behaviour and the attitudes that one holds. It is the extent of this conflict that determines the degree of dissonance that is aroused. This, in turn, will depend upon the conditions governing the behavioural change itself.

Individuals may engage in behaviour which is inconsistent with their attitudes for a variety of reasons. In some cases this may occur 'by accident' as envisaged in Hirschman's case where individuals 'stumble more or less absent-mindedly into such behavior'. In most cases, however, the discrepant behaviour will occur as a result of inducements or outright coercion (as when the behaviour is made compulsory by government). These different types of motivation are important because they have quite different implications for the amount of dissonance that is likely to be aroused.

Consider first the extreme case where the altered behaviour is entirely due to coercion (which is known in the dissonance literature as the case of 'forced compliance'). Here little or no dissonance will result since the behaviour change can be easily dismissed as something that 'had to be done' and for which the individual consequently bears no responsibility.[8]

A similar argument applies to the case where behavioural change is induced not by outright coercion, but rather by some form of incentive scheme. In general, the theory predicts that the greater the incentives under which the behavioural change takes place the less will be the dissonance and consequently also the degree of attitude change. For the greater are the incentives the more easily can the behaviour change be considered (in a way that is analogous to forced compliance) as something that was done because of the circumstances rather through an act of individual volition. And lacking the volitional component, the behaviour change fails to produce dissonance or a change in attitudes.

Firms trying to promote the sales of new products through free samples, cut-price offers, etc., frequently encounter this type of relationship between attitudes and behavioural change. Thus the more incentives are used to induce buyers to change their behaviour patterns (i.e. to switch to the new brand) the less likely is it that attitude change in favour of the product (and hence repeat purchasing) will ensue.[9] Indeed, from this point of view, it may be preferable to charge something than offer free samples. And in the forced compliance literature numerous experiments have shown that the greater the inducement to make a public statement that is contrary to one's private opinion, the less is the resulting change in attitude in the direction of the view expressed in the public statement.

What the above discussion suggests as a necessary condition for dissonance arousal, therefore, is that the individual experiences some degree of choice in making his or her change of behaviour. Even if this condition is met, and dissonance is therefore aroused, however, it does not necessarily follow that the outcome will be a change in attitude. For what is then crucial is how the individual chooses to reduce the dissonance that is aroused.

The mode of dissonance reduction

'Dissonance theory is nonspecific with regard to the mode of dissonance reduction used by the individual after dissonance arousal.'[10] That is, it makes no specific prediction as to whether dissonance will be reduced by one means rather than another. What the theory does say, however, is that in general the individual will opt for the mode of dissonance reduction that offers the least resistance.

Consider the example of the smoker who experiences dissonance because the act of smoking is inconsistent with his knowledge that it is bad for his health. Because it is very difficult for him to deny that the habit is bad for his health, he may instead attempt to reduce dissonance by arguing that if he stopped smoking he would put on weight which is equally bad for his health.

It follows that if attitude change is to follow discrepant behaviour change as the chosen mode of reducing the dissonance occasioned by the latter, it must offer less resistance than other possible modes, perhaps the most obvious of which is attempting to revert to behaviour that is more in line with one's attitudes. The question is then whether it is easier to alter one's new behaviour or one's old attitudes. Among the crucial determinants of this choice are, first, the degree of irrevocability of the former, e.g. whether or not the individual would incur sizeable financial losses from 'uncommitting' himself from his new behavioural commitment. The difficulty of changing one's attitudes – the second major determinant – depends, on the one hand, on how entrenched these have become, and on the other hand, on how much social support one has in changing them.[11]

IMPLICATIONS FOR CONSUMPTION POLICY

In the previous section we examined the conditions under which a change in behaviour produces an accommodating attitude change. Brehm and Cohen have summed up the implications of these conditions for social policy in the following way:

> The strongest effects on attitudes . . . and subsequent behavior are produced when careful thought is given to the conditions under which people are forced to comply with the demand that they do something they do not want to do. Changes in attitudes, and so forth, away from an initial position one has held, toward a new and originally unpleasant position, depend certainly on first arranging the social conditions that elicit the new and recalcitrant response. However, *it is the manner in which the environment is organized in getting the person into the new situation, or arranged once he is in it, that determines how well he will adjust to it by accepting it, incorporating it, rationalizing it, and, finally, even proselytizing for it.*[12]

Because many, if not most, development policies are designed eventually to alter the behaviour (including consumption-oriented behaviour) of certain members of poor societies in ways that conflict with the (often traditional) attitudes held by these individuals, we believe that this statement has relevance also for this area of policy-making. Through an interpretation of some of these policies in the light of the analysis in the previous section, we shall therefore attempt to throw light on their frequently observed failure (as well as occasional successes) to produce accommodating attitude change, and in this way to indicate the direction future policies might take in order to raise their chances of success in this respect.

The above analysis of the necessary conditions for accommodating attitude change resultant upon behaviour change suggests two major problems for development planners. The first is how to induce a change in behaviour without at the same time negating the conditions for the arousal of dissonance. Thus, 'While it is necessary to have a minimum of . . . extrinsic motivations in order to obtain compliance . . . the strength of any of these forces beyond the minimum makes a great deal of difference in ultimate attitude change.'[13] In other words, what policy-makers confront is a vicious circle; getting people to engage in dissonance-arousing activities would seem to require coercion or strong incentives but the very use of these methods defeats the objective of engendering sufficient dissonance to produce a change in attitudes. (This difficulty, it bears emphasizing, will be all the more acute when fundamental human values are at stake and when these values are deeply entrenched.)

One way for policy to overcome this 'Catch 22' is to make use of group pressure, which, under some circumstances, may appear as less coercive than pressure exerted directly on individuals from above. These favourable circumstances seem to arise when the individual feels part of a cohesive group, whose norms he (therefore) likes to live up to and in whose decisions he takes some active role.[14] Group change will then tend to be easier to effect than change of individuals considered separately. During the Second World War, for example, psychologists who were seeking the best way to convince housewives to buy unfamiliar types of foods found that 'More housewives changed their behavior and attitudes about various types of foods after participating in a group discussion than after hearing a lecture on the value of eating these same new foods. Once the new group standard had been formed through group discussion, it was easier for the housewives to change their behavior to conform to the new standard.'[15]

Much the same use of peer pressure in small groups (*hsiao-tsu*) appears to have been made for the purpose of mobilizing the Chinese population during the Maoist period. But as Whyte has pointed out, effective use of this type of pressure demanded an extremely subtle blend of psychological forces, which was not at all easy to establish. In particular:

> Mutual criticism in a sense is the prime motive force behind attitude change in *hsiao-tsu*; as such it was extremely important that mutual criticism and the atmosphere surrounding it be correct. The desired atmosphere was one quite different from that valued in traditional Chinese primary groups. The terms used convey this point. What was desired was a 'happy, tense, lively atmosphere', . . . 'solidarity but not harmony' . . . Thus a new kind of

> unity was to develop, one based on rigorous group criticism rather than on the smoothing over of conflict in order to preserve harmony (the traditional emphasis). On the other hand, *hsiao-tsu* were not supposed to be the scene of relentless and merciless criticism, which would serve only to alienate members. Solidarity and criticism were to be combined, with criticism leading not to disharmony but to improved unity . . . As those criticized acknowledged their defects and resolved to change their ways, unity was restored, but at a higher level (i.e., with the defects corrected). *Only if criticism was properly balanced with group solidarity could it be effective in changing the attitudes of members.*[16]

A second way in which policy may be able to overcome the vicious circle described above is through use of the so-called 'foot-in-the-door technique'. This technique of producing compliance with a minimal amount of apparent pressure is based on the assumption that 'once a person has been induced to comply with a small request he is more likely to comply with a larger demand.'[17] It is not entirely clear why this technique appears to have so powerful an effect. The most likely explanation, according to Freedman, Sears and Carlsmith,[18] is that 'people who agree to a small request get involved and committed to either the issue itself, the behavior they perform, or perhaps simply the idea of taking some kind of action. Any of these involvements would probably make someone more likely to comply with future requests.' One of the most interesting and widespread applications of the foot-in-the-door approach is to be found in sales campaigns that attempt to engage the consumer in even the most minor forms of initial involvement with the product (such as sending back a card indicating that he does not want the product).

Even if policy-makers are able successfully to employ techniques of peer pressure and gradual persuasion to bring about temporary counter-attitudinal behaviour, they will still, however, face the problem of ensuring that this behaviour *persists* and for this more sustained behavioural alteration a fundamental change in attitudes is generally likely to be necessary. The point can usefully be illustrated with reference to the distinction that Brehm and Cohen draw between what is required for racial desegregation on the one hand and more fundamental racial integration on the other. As they put it, 'It is true that many of the advocates of desegregation feel that it does not matter *what* people's attitudes are or *how* they react so long as desegregation, and therefore racial equality, is brought about. But, if one is oriented toward the long view, and takes the position that desegregation is after all the first step toward integration, he should examine thoroughly the conditions under which desegregation takes place.'[19] Desegregation which occurs, for example, in the context of an ambiguous public policy, ineffective police behaviour or conflicts between public officials themselves, tends to produce violence and a strenghthening of initially unfavourable values rather than any movement towards values that are consistent with racial integration.

Brehm and Cohen also draw attention to the social pressures that form an important part of the context in which the individual decides whether or not to use value change as a means of reducing the dissonance to which his involvement in desegregation gives rise. 'Thus positive attitude change may take place provided that the cost of such a cognitive change (measured in terms, for example, of social disapproval or ostracism) is not too severe for the individual.'[20] This appears to have been the case in desegregation which took place in the context of public housing schemes, largely it seems, on account of the commitment made by participants to the explicit goals of these schemes. There is an interesting contrast here with private housing which allows people to come and go as they please and which involves no commitment to an 'overt policy' of racial interaction. In this type of situation, value change is unlikely to constitute the preferred mode of dissonance reduction. Indeed, 'even those with favorable attitudes may leave, simply because of response to social pressures and norms'.[21] Thus, according to Brehm and Cohen, 'public housing which requires some physical desegregation will more certainly be a locus for dissonance arousal and reduction via favorable attitude change . . . than will private housing.'[22]

In general, policy-makers need to ensure that attitude change is the preferred (i.e. least-cost) mode of dissonance reduction (assuming, as we are, that the change in behaviour has been accompanied by a relatively high degree of dissonance). Policies for this can take the form either of making 'uncommitment' of behaviour difficult, or of facilitating accommodating attitude change, or both of these. Achievement of the former can be promoted by ensuring that the policy underlying the induced behaviour is firm and irrevocable and the latter by mobilizing village leaders or other authority figures to sanction the accommodating attitude change, for it is a fundamental general proposition of dissonance theory that 'dissonance will be resolved via positive attitude change . . . *to the degree that such positive attitudes are socially sanctioned*.'[23]

How successfully the actual practice of development has dealt with these problems will be the subject of the following section, in which we present case-study material from a variety of different areas of policy-making, with particular reference to consumption-oriented development policies.

POLICY CASE STUDIES

Population policy

For a time, it seemed reasonable to those concerned with policies for family planning that initial acceptance of family planning methods meant satisfaction and, furthermore, that initial acceptance and other short-time indicators were closely associated with long-term use.[24]

Subsequent findings, however, belied the view that success in family planning methods could be usefully measured by acceptor rates. In particular, it was found that continuation rates 'were extremely low in studies from several dozen selected countries.'[25]

These later findings are consistent with the dissonance approach, which, in general, predicts that a change in behaviour (in this case acceptance of contraceptive methods) *may or may not* be associated with a corresponding attitude change and that there is no necessity for acceptor rates to be highly correlated with long-run success.[26] According to the theory, the change in attitude – and consequently the degree of continuation – depends crucially on the *precise manner* in which the initial change in behaviour is effected.

Consider the case of the individual with an initially unfavourable attitude towards family planning. The problem is not only to secure behaviour on the part of the individual that is inconsistent with his/her initial attitude, but also to secure a *sustained* change in behaviour which, in turn, demands a change in attitude. A major problem is that even if the first hurdle can be overcome and the initial behaviour change is achieved (so that dissonance is aroused), it can all too easily be dispelled by reversion to one's initial behaviour. That is, without some degree of commitment to the new mode of behaviour, behaviour rather than attitude change is likely to be the preferred mode of dissonance reduction.

An example from a contraceptive distribution scheme in Bangladesh illustrates the problems. In 1975 all married women of reproductive age were given information on and offered a supply of oral contraceptives. If they accepted, they were given a supply of six cycles. However, the women were not asked to commit themselves to using the contraceptives. Though contraceptive use rose quite sharply during the following year, the dropout rate was reported to be high.[27] At least part of the reason for this may have been the lack of commitment – and the consequent ease of using 'uncommitment', rather than attitude change, as a mode of dissonance reduction.

Similar problems of securing sustained behavioural change are often encountered in policies for compulsory education. With respect to the Chilean case, for example, Anderson and Bowman report that despite the existence of legal compulsory primary education, 'There is a very heavy rate of school desertion . . . in direct violation of the law.'[28] With poor enforcement of the law, it is all too easy to reduce dissonance-arousing behaviour by changing the behaviour itself.

As noted in the previous section, attitude change may also fail to occur because of the way in which coercive-type methods are used to bring about the initial change in behaviour. This problem, too, can be illustrated in the context of incentive policies for family planning.

Many developing countries have experimented with incentives of various kinds in the promotion of family planning practices. In some cases, incentives are provided to those responsible for promoting birth control in the field, who are often referred to as canvassers. This had led, on occasion, to results quite contrary to the objectives of the programmes. In particular, canvassers have used a high degree of coercion to secure behaviour change (e.g. through vasectomies). As a result, respondents have *continued to hold an unfavourable attitude towards vasectomy even after adoption*. And, 'Such adopters are likely to spread unfavorable rumors about vasectomy. The result is an eventual plateau effect in the rate of adoption . . . A dissatisfied customer who adopted because of pressure from a canvasser is the worst kind of interpersonal advertisement, for an innovation.'[29] (Note here that a similar outcome may be predicted by dissonance theory if, without actual coercion, incentives are very high.) More generally, there is 'sufficient experience with disastrous results of forced family-planning programs to support the proposition that coercive compliance structures are inappropriate, and may be counterproductive, for the goals of population planning.'[30]

Rogers concludes his analysis of the efficacy of incentives for family planning with a plea for enriched 'understanding of the sociopsychological process of persuasion and innovation decision making through which incentives have an effect. How individuals perceive incentives and how such perceptions structure their motivation and affect their decisions are not yet known.'[31]

All of the above has been concerned with the second main question noted at the start of this section, namely whether or not attitude change accompanies behavioural change (which assumes that the latter has actually been secured). But in the field of policies for family planning, securing the initial desired behaviour change has often proved very difficult. Indeed, from field experience it is clear that changing values, on the one hand, or making 'structural changes' on the other, is not always enough to change behaviour.[32] That is, while non-coercive methods of securing behaviour change have often failed, methods that are 'too coercive' also, as we have seen, may fail to achieve the attitude changes that are required for a sustained improvement in methods of birth control.

Motivational aspects of the successful Chinese case

In the above, we argued that the dissonance formulation is able to throw light on certain instances of failure in the field of family planning. We also believe that the theory provides a useful framework within which aspects of the successful Chinese experience in this area can be understood.

As Leo Orleans has pointed out, the Chinese effort to introduce birth control to the population has had to overcome virtually all the problems that are typically encountered by other developing countries, namely, 'traditional values with regard to family, the desire for children . . . an attitude toward sex that inhibited open discussion of the subject, the traditional subservient role of the woman'.[33] He also indicates that with respect to the delivery system that the Chinese used to solve these problems, there is little that is new to family planning specialists. What does distinguish the Chinese approach appears to be its motivational aspect and we shall offer a dissonance theory interpretation of this experience (while at the same time noting that this is not the only possible interpretation).

The essence of the motivational system that is applied to population control is peer pressure. This pressure is felt both at the workplace and at the residential level. With respect to the former:

> The involvement of the management and of the workers and employees or peasants in the personal lives of each individual extends considerably beyond work-related subjects . . . study

topics related to family planning are common . . . It is not unusual for the women workers in a factory, for example, to announce their joint decision as to the number of births they will collectively have during the following year, or to see a chart on the wall of a hospital or an office.[34]

The focus of peer pressure at the residential level is the basic level cadre whose function is to personally guide and continually assist mothers and potential mothers. In some cases, the degree of involvement of the local community in the choice of family size is such that debates are held to discuss which families would have children during forthcoming years. As a consequence of these powerful pressures that have been brought to bear on the Chinese population it used to be the case that 'a couple that has too many children not only rejects Mao's wishes but disgraces its own group or community.'[35]

One of the striking aspects of the way in which the Chinese confronted the intractable problem of 'traditional attitudes' in birth control as well as in other areas during the Maoist period is that it appears to conform quite closely to what the theory described in the previous section would require for favourable attitude change to accompany prior alterations in behaviour. Indeed, as Whyte[36] points out, much of the reasoning presented in political pamphlets parallels closely the findings of Western social psychologists, although the pamphlets do not cite these findings and do not discuss the issues in the same terms. In particular, the Chinese laid great stress on the need to avoid behavioural change through forced compliance: 'Indirect pressure, persuasion, and the force of example should be emphasized and commands and threats deemphasized.'[37] That is, behavioural change was to be brought about in a manner that would arouse dissonance on the part of those whose initial attitudes were inconsistent with the change. For this purpose, as noted earlier, the Chinese leaders relied heavily on the peer pressure that was exerted through the small group, achieving in this way, what often appears as paradoxical about the Maoist development model, namely the combination of regimentation and mass enthusiasm. 'To the Western mind (these) seem quite contradictory, or attainable only in very special organizations with highly selected participants (perhaps in a monastery).'[38]

But it is important to note that in the use of this method the Chinese were not always successful in achieving the ideal type of small group structure which was described in the theoretical section above and through which compliance could be secured with a minimum of apparent pressure. As Whyte has again pointed out:

> In some organizations individuals do feel pressure from peers whom they respect to change their attitudes, and find the prospect of 'falling behind' hard to bear. But in some other organizational settings individuals feel they are surrounded by opportunists waiting to criticize their slightest misstep in order to gain favor with superiors. In other settings individuals look upon political rituals as a bothersome but unavoidable routine, and save their energies and emotions for activities outside the group. In these and other ways the ideal strict political atmosphere is not achieved, and the psychological impact of group activities is reduced, or even reversed.[39]

In terms of dissonance theory, all these difficulties may be said to have the effect of frustrating the attempt to produce *sustained* behavioural change through attitude change that is (intended to be) produced via (non-coercive) peer pressure in the small groups. And as a consequence, 'In sustaining high commitments and efforts, the elite still relies upon material incentives and sanctions more than it would apparently like to.'

The Indonesian population programme

Like its Chinese counterpart the Indonesian population programme is often thought of as being highly successful; according to one commentator for example, 'There is considerable evidence today that the Indonesian Population Program has done outstandingly well',[40] regardless of whether it is judged according to the numbers of contraceptive users that have been reached or according to its impact on fertility reduction. And like the Chinese case, the successful outcome of the Indonesian programme seems at least in part, to be attributable to factors that can plausibly be interpreted in terms of cognitive dissonance theory.

Most importantly, there has been a reliance on voluntary acceptance of family planning rather than on material incentives or on methods of coercion. The former were used at one time for family planning acceptors but were found to be ineffective. (Interestingly, in relation to staff motivation as well, 'the extent to which direct monetary rewards have been used . . . to stimulate performance has been limited.'[41]) And while there have been allegations that coercive methods were used to influence the population (as part of 'special drive' tactics):

> Surveys . . . have shown that where the program was able to mobilise a wide range of formal and informal leaders in motivation and information, people felt, by far, that their acceptance was voluntary. Where there was little participation by village leaders and the headman merely transmitted the program aims, the message came to be perceived as an order backed by force . . . On the basis of field evidence and other reports, many experts have concluded that the allegation of coercion is exaggerated and that the Indonesian program's success is, on the whole, *due to voluntary acceptance and participation*.[42]

Again, as in China, this achievement appears to be closely related to the mobilization of powerful peer pressure at the level of the small group. There is unfortunately not much detailed evidence of how this process actually worked, but it does seem at least superficially to resemble the Chinese small-group model that was described above. Consider from this point of view the following description of the village-level family planning programme in Indonesia that Samuel Paul has provided:

> During the monthly meetings of the male heads of households, community issues such as taxes, irrigation, disputes and festivals used to be discussed and resolved through group consensus. Now family planning was also added as an additional topic for community deliberation and increasing responsibility was placed on the hamlet's administrative structure for managing its own family planning services. Each head of a household, for example, was required to report publicly the family planning and pregnancy status of every married woman in his household. Banjar registers were compiled listing all eligible couples, their location, use of contraception, and information of their receipt of pill and condom resupplies . . . Maps of the banjar were drawn which identified each household according to its eligibility for and use of contraception by method, and these were prominently displayed in the banjar meeting hall.[43]

In so far as those who were thus influenced by the family planning programme did indeed regard their compliance with it as essentially voluntary, one would expect (for reasons that have already been given) cognitive dissonance to have been aroused among them. Another aspect of the programme, moreover – the involvement of important religious leaders – may have played a role in increasing the likelihood that value change was the method used to reduce the dissonance thus aroused. For in a predominantly Islamic country such as Indonesia, the disapproval of family planning by religious leaders would have made it all but impossible to provide the social sanctioning that appears to be so central to the process of value change according to cognitive dissonance theory. Those who designed the Indonesian programme managed to avoid this problem by the careful and deliberate persuasion and involvement of Islamic leaders at an early stage; in 1972 for example, at a well-publicized conference of some of these leaders, approval of family planning was announced.[44]

The diffusion of sanitation technology in Mexico

Several of the elements that characterize the general Chinese and Indonesian models can also be studied in operation at the micro-level of the successful introduction of new sanitation technology in Yalcuc, Mexico.

Latrines were first introduced into Yalcuc about 20 years ago and the subsequent experience in that community, 'is an example of rapid diffusion and adoption of an innovation in a harmonious manner beneficial to the majority.'[45] In particular, following an initial adoption rate of almost 90 per cent, the rate of continuous use over the ensuing 20-year period was 75 per cent.

Much of the enduring success of this project appears to have been due to the precise manner in which it was first implemented. Specifically:

> the decision to install latrines was made within the context of a community project; all the aura of the leadership and the pressures of social control were brought to bear on the villagers. Gradually, *all the men in the community signed an agreement in a village meeting signifying their commitment to the collective decision to install the latrines* . . . The decision to install latrines was as much a matter of community dynamics as of individual decision-making . . . The costs (in money and time) of installing a latrine were perceived by many as minor compared to the costs (in social pressure, loss of good will, and deterioration of solidarity) of *not* installing one.[46]

In contrast to the contraceptive distribution scheme in Bangladesh referred to above (where the initial change in behaviour was not sustained), what appears thus to have occurred in Yalcuc was that the commitment to the project that was jointly enforced through group pressure and in writing caused attitude change to be the preferred mode of dissonance reduction. Peer pressure in this case, as in the more successful Chinese and Indonesian examples, appears to have been perceived by participants as something much less coercive than a mere directive imposed from above. Indeed, the group structure that was employed for the purpose of diffusion of the latrines resembles in key respects the 'ideal-type' sought by the Chinese leaders. In particular, 'Proposals are discussed thoroughly, and opposing opinions are heard respectfully . . . Never is a vote taken, and as a result, polarization is minimized.'[47] And in this way, 'The concern with solidarity is supported by a system of social pressure to comply with collective decisions and generally follows the norms of the community.'[48] Taken together, these aspects of the way in which the project was implemented helped to ensure a lasting alteration in behaviour, as was manifest in the long-term continued use of the latrine.

If, therefore, the Yalcuc sanitation project provides an illustration of the way in which peer pressure can be used to confront the problem of securing counter-attitudinal behaviour with a minimal amount of perceived coercion, other apparently successful projects in this field suggest that attention should also be paid to the potential of the alternative 'foot-in-the-door' (i.e. gradual persuasion) technique. In two small communities in rural Guatemala, for example, the initiators of a potable water and latrines project managed first 'to gain the allegiance of a small group active in the community and capable of generating community support. The commitment of the rest of the community members was obtained in a gradual process of individual persuasion by core leaders.'[49]

IS COGNITIVE DISSONANCE A UNIVERSAL PHENOMENON?

As noted above, cognitive dissonance theory is based on the assumption that there is a tendency towards cognitive consistency (that is manifest in attempts to reduce or eliminate cognitive inconsistencies or dissonance). Because the empirical verification of this theory has been conducted almost entirely in developed countries, our extension of it to the Third World context effectively assumes that the need for cognitive consistency is a universal one.

Certainly, Festinger, the inventor of the theory seemed to think that this was the case; indeed, as he expressed it in his original formulation, the human need for cognitive consistency is capable of being placed on the same footing as the basic human instincts for hunger, thirst, etc.[50] (A similar question, it is interesting to note, arises in relation to the universality of major elements of Freudian psychoanalytic theory. In this regard, Erich Fromm has argued that the 'instinctual apparatus, both quantitatively and qualitatively, has certain biologically determined limits to its modifiability and that only within these limits is it subject to the influence of social factors.'[51])

On the other hand, it is sometimes suggested that the human need for cognitive consistency is far from universal and that in at least one developing country, India, a distinct tolerance for inconsistency can be observed. One can find, for example, references to how apparent 'contradictions of thoughts and actions, instead of leading to confrontation, are tolerated, balanced, accommodated and integrated . . . It is this unique tolerance of contradictions which sometimes appears paradoxical in the Indian psyche.'[52] For other observers, the tolerance of contradictions (or dissonance) is in fact 'a major attribute of the Indian Self'.[53]

Pepitone and Triandis have questioned the universality of cognitive dissonance theory from the wider perspective of 'cultures where beliefs in mystical causation are pervasive' and where in consequence 'inconsistency may not be a definable state between cognitions'.[54]

Such queries as these form part of a larger debate that – in a manner analogous to the way in which development economists tend to take issue with the notion of a 'monoeconomics' – has challenged the applicability of psychological theories developed in and for the industrialized countries to the Third World. Jahoda, for example, has suggested that:

> the theories prevalent in experimental social psychology are the product of a specific social milieu, namely advanced industrial societies. Their crucial attributes from the present standpoint are those of literacy, impersonality (i.e. in sociological jargon they are universalistic as opposed to particularistic), and a *wide range* of available beliefs, ideas and attitudes. The joint effect of all these is to permit individuals a substantial area of free social space. In traditional communities such space is far more restricted, in terms

> of beliefs and attitudes as well as overt behaviour. It is not merely that there are practically no 'strangers' within the social system but relationships to all kinds of others, including kin, tend to be more closely regulated.[55]

As yet, however, it is largely unclear how and to what extent these and other differences between rich and poor countries bear on the validity in the latter of social psychological theories developed for the former. For, as Amir and Sharon have pointed out, it is still very much the case that most studies neglect the cross-cultural dimension and 'assume global validity to their findings'.[56] In fact, according to one estimate, less than ten per cent of all empirical research in social psychology 'strategically involves samples drawn from two or more cultures'.[57] It thus remains largely to be seen if social psychological theories in general and cognitive dissonance theory in particular retain their validity when they are transferred from the context of the industrialized countries to the very different circumstances prevailing in the Third World.

CONCLUSIONS

Over a wide range of consumption-oriented activities[58] in developing countries, such as health, nutrition and family planning, governments are concerned to alter norms. In this endeavour, they will often have to confront deep-seated and unyielding values, about which a great deal has been written in sociology and psychology. In much of this literature, it is assumed that changes in these values are a necessary prerequisite for the desired changes in behaviour (for example, with respect to the adoption of methods of family planning).

Almost twenty years ago, however, Hirschman pointed out that this sequence may not in fact be necessary, and indeed, that according to the social psychological theory of cognitive dissonance the reverse may occur (that is, favourable attitude change may be the result rather than the cause of alterations in behaviour). Because Hirschman did not, however, deal with the conditions that are required for this reversal of the traditional sequence to occur, we have tried to extend his analysis to incorporate a specification of the conditions laid down by dissonance theory. In addition, we have sought to show that numerous instances of success and failure in policies designed to promote attitude change in the Third World can plausibly be interpreted in the light of these conditions. To the extent that our interpretations are valid, a number of powerful implications for the design of consumption policy emerge from the analysis. To be more specific, policy-makers will have to pay especially close attention to the subtle psychological requirements for bringing about counter-attitudinal behaviour with a minimum amount of perceived pressure.

In assessing the validity of our interpretations, the reader should recognize that underlying these is the assumption that cognitive dissonance theory is 'universally correct'. That is to say, we have implicitly assumed not only that the theory is applicable to the modern Western type of society for which it was originally intended, but also that it is largely transferable to the often traditional societies of the Third World. With respect to the former, we can do no better than refer the reader to the specialized experimental evidence.[59] The case for the transferability of the theory seems to us to depend partly on whether the human need for cognitive consistency can really be placed on the same footing as the basic human instincts for hunger, thirst, etc. (as Festinger suggested in his original formulation), and if so, on the extent to which even these drives are capable of being modified by different environments. As yet, however, very little evidence is available to assess the extent to which social psychological theories in general and cognitive dissonance in particular retain their validity when they are transferred from rich to poor countries.

Notes

1. As Behrman (1989), p. 1518 has aptly put it, the fact 'that efforts to change norms seem fairly widespread in the real world suggests, by a revealed preference argument, that some people believe that the resources devoted to such activities are well-utilized which presumably means effective in changing norms.'
2. According to Blackler (1983), for example, 'while Western anthropologists, sociologists, geographers, economists, and political scientists can point to important achievements with regard to an analysis of problems faced by developing countries, social psychological contributions are far less in evidence' (p. ix).
3. Hirschman (1965, 1970, 1971, 1981 and 1982).
4. Hirschman (1965).
5. Ibid., p. 392.
6. Ibid., pp. 392–3, emphasis added.
7. Brehm and Cohen (1962), p. 272.

8. Freedman, Sears and Carlsmith (1978) refer in this context to the ease with which behaviour can be changed in comparison with changes in attitudes. 'There are tried and true techniques of social influence (or other forms of power) that can produce behavior wholly contradictory to the person's attitudes, or even his perceptions of reality. Subjects freely administer shock to inoffensive people when they are told to, or express beliefs that wholly contradict their sense perceptions. In considering these dramatic cases it is well to contrast the ease of behavior change with the manifest difficulty of attitude change, at least of attitudes that matter' (pp. 381–2).
9. Wicklund and Brehm (1976).
10. Brehm and Cohen (1962), p. 280.
11. Hirschman (1970), pp. 146–7 attempts to identify the conditions for different responses to dissonant situations on the part of group members. The dissonance aroused is a consequence of the difference between the 'cognition that (one) has undergone an unpleasant ordeal to attain membership and the cognition concerning the undesirable aspects of the group. As a mode of dissonance reduction, 'exit' is likely to occur when the costs of exit are low, which may be partly a consequence of the lower dissonance that is aroused by a 'mild initiation to the group'.
12. Brehm and Cohen (1962), p. 298, emphasis added.
13. Ibid., p. 283.
14. Hare (1976).
15. Ibid., p. 53.
16. Whyte (1974), p. 49, emphasis added.
17. Freedman and Fraser (1966), p. 195.
18. Freedman, Sears and Carlsmith (1978), p. 418.
19. Brehm and Cohen (1962), pp. 284–5.
20. Ibid., p. 280.
21. Ibid., p. 281.
22. Ibid., p. 279.
23. Ibid., p. 284, emphasis added.
24. Bruce (1980).
25. Zeidenstein (1980).
26. Though of course there are other possible reasons for these findings.
27. Huber and Khan (1979).
28. Anderson and Bowman (1971), p. 305.
29. Rogers (1971), p. 246.
30. Ness (1979), p. 636.
31. Rogers (1971), p. 247.
32. Knight (1980).
33. Orleans (1975), p. 503.
34. Ibid., p. 515.
35. Ibid., p. 525.
36. Whyte (1974).
37. Ibid., p. 53.
38. Ibid., p. 17.
39. Ibid., p. 231.
40. Paul (1982), p. 63.

41. Ibid., p. 76.
42. Ibid., p. 77, emphasis added.
43. Ibid., p. 68.
44. Ibid.
45. Elmendorf and Buckles (1980), p. 20.
46. Ibid., pp. 19–20, emphasis added.
47. Ibid., p. 20.
48. Ibid.
49. Ibid., p. 14.
50. Festinger (1957).
51. Fromm (1978), p. 163.
52. Sinha (1988), p. 44.
53. Pepitone and Triandis (1987), p. 491.
54. Ibid., p. 482.
55. Jahoda (1983), p. 23, emphasis in original.
56. Amir and Sharon (1987), p. 386.
57. Pepitone and Triandis (1987), p. 476.
58. Cognitive dissonance theory can be applied to other areas of development as well. It can help, for example, to explain the failure of the Ujamaa village scheme in Tanzania. What appears to have been lacking in that scheme, among other things, was the operation of any mechanism that provided social pressure of the Chinese small-group type. 'What was needed was a mobilizing agency which could mediate between the goals determined by the state and the perhaps understandable but nevertheless regrettable reluctance of the peasant' (Turok, 1975, p. 415).

    Dissonance theory can also be applied to the behaviour and attitudes of the government itself. Hirschman, for example, has suggested that the operation of the forced compliance principle makes it difficult to use foreign aid as an effective means of influencing change on the part of unwilling aid recipients (Hirschman, 1965). A very similar argument can be made in the case of IMF conditionality which does not often seem to exert any lasting influence over recipients. One commentator in fact has noted that 'even where governments accept the economic wisdom of the Fund prescriptions, it may prove politically more convenient for them to present the policies as "Fund policies" which have been unwillingly forced on them' (Bird, 1984, p. 176).
59. See, for example, Wicklund and Brehm (1976) and Freedman, Sears and Carlsmith (1978).

References

Amir, Y. and Sharon, I. (1987). 'Are Social Psychological Laws Cross-Culturally Valid?', *Journal of Cross-Cultural Psychology*, Vol. 18 (4).

Anderson, C. Arnold and Bowman, M.J. (eds) (1971). *Education and Economic Development*, Aldine.

Behrman, J. (1989). 'The Simple Analytics of Contraceptive Social Marketing', *World Development*, Vol. 17 (10).

Bird, G. (1984). 'Relationships, Resource Use and the Conditionality Debate', in Killick, T. (ed.), *The Quest for Economic Stabilization*, Heinemann.

Blackler, F. (1983). 'Preface: The Plan of the Book', in Blackler, F. (ed.), *Social Psychology and Developing Countries*, Wiley.

Brehm, J.W. and Cohen, A.R. (1962). *Explorations in Cognitive Dissonance*, Wiley.

Bruce, J. (1980). 'Implementing the User Perspective', *Studies in Family Planning*, Vol. 11 (1).

Elmendorf, M. and Buckles, P. (1980). *Appropriate Technology for Water Supply and Sanitation: Sociocultural Aspects of Water Supply and Excreta Disposal*, The World Bank, Washington DC.

Festinger, L. (1957). *A Theory of Cognitive Dissonance*, Stanford University Press.

Freedman, J.L. and Fraser, S.C. (1966). 'Compliance Without Pressure', *Journal of Personality and Social Psychology*, Vol. 4 (2).

Freedman, J.L., Sears, D.O. and Carlsmith, J. Merrill (1978). *Social Psychology*, 3rd edn, Prentice-Hall.

Fromm, E. (1978). *The Crisis of Psychoanalysis*, Pelican.

Hare, A. Paul (1976). *Handbook of Small Group Research*, 2nd edn, Free Press.

Hirschman, A. (1965). 'Obstacles to Development: A Classification and a Quasi-Vanishing Act', *Economic Development and Cultural Change*, Vol. 13 (4).

Hirschman, A. (1970). *Exit, Voice and Loyalty*, Harvard University Press.

Hirschman, A. (1971). *A Bias for Hope*, Yale University Press.

Hirschman, A. (1981). *Essays in Trespassing*, Cambridge University Press.

Hirschman, A. (1982). *Shifting Involvements*, Princeton University Press.

Huber, D. and Khan, A.R. (1979) 'Contraceptive Distribution in Bangladesh', *Studies in Family Planning*, Vol. 10 (8/9).

Jahoda, G. (1983). 'Has Social Psychology a Distinctive Contribution to Make?', in Blackler, F. (ed.), *Social Psychology in Developing Countries*, Wiley.

Knight, P. (1980). *Implementing Programs of Human Development*, World Bank Working Paper No. 403.

Ness, G.D. (1979). 'Organizational Issues in International Population Assistance', in Hauser, P. (ed.), *World Population and Development: Challenges and Prospects*, Syracuse University Press.

Orleans, L. (1975). 'China's Experience in Population Control: The Elusive Model', *World Development*, Vol. 3 (7 and 8).

Paul, S. (1982). *Managing Development Programs: The Lessons of Success*, Westview.

Pepitone, A. and Triandis, H.C. (1987). 'On the Universality of Social Psychological Theories', *Journal of Cross-Cultural Psychology*, Vol. 18 (4).

Rogers, E. (1971). 'Incentives in the Diffusion of Family Planning', *Studies in Family Planning*, Vol. 2 (12).

Sinha, D. (1988). 'Basic Indian Values and Behaviour Dispositions in the Context of National Development: An Appraisal', in Sinha, D. and Rao,

H. (eds), *Social Values and Development*, Sage.

Turok, B. (1975). 'The Problem of Agency in Tanzania's Rural Development', in Cliffe, L. *et al.* (eds), *Rural Cooperation in Tanzania*, Tanzania Publishing House.

Whyte, M.K. (1974). *Small Groups and Political Rituals in China*, University of California Press.

Wicklund, R.A. and Brehm, J.W. (1976). *Perspectives on Cognitive Dissonance*, Lawrence Erlbaum.

Zeidenstein, G. (1980). 'The User Perspective: An Evolutionary Step in Contraceptive Service Programs', *Studies in Family Planning*, Vol. 11 (1).

11 Income Redistribution, Consumption and Employment*

INTRODUCTION

The ILO Mission Report on Colombia espoused the idea that since the poor tend to consume relatively more labour-intensive commodities than the rich, a redistribution in their favour will tend to increase employment.[1] This proposal led subsequently to a host of attempts to demonstrate its empirical validity at the economy-wide level in a number of less-developed countries.[2] The results, however, gave only qualified support to the hypothesis. As Morawetz, put it, 'The almost unanimous conclusion is that even quite significant redistributions of income seem likely to have only marginal effects on growth and employment, usually increasing the latter by less than 5%.'[3]

This chapter is concerned to examine one possible reason for these disappointing results, namely the aggregation of products into broad, usually two-digit categories. Aggregation in this sense involves an estimation bias if there are shifts *within* these categories to goods produced by different factor combinations. In this study we attempt to examine the nature and extent of aggregation bias within a particular industry – that of sugar processing in India.

PRODUCTS AND TECHNOLOGIES IN THE INDUSTRY

Sugar processing in India is a good example of an industry in which two similar but distinct products serve the different ends of the market. A traditional product, gur, which is basically sugarcane in concentrated form is still the dominant sweetener for the vast majority of those with low incomes, particularly in rural areas. Some 60 per cent of the annual sugarcane crop is used for gur manufacture and 28 per cent of the remainder goes to the refined sugar industry. Annual

* First published by the International Labour Office in the World Employment Programme Research Working Paper Series, November 1977.

per capita consumption of gur is roughly 12 kilos as opposed to 5 kilos in the case of sugar.[4] Gur represents on average 1.5 per cent of total consumption expenditure in rural areas and 0.6 per cent in urban. The consumption patterns in the industry can be represented as in Table 11.1 which shows the data presented by monthly per capita expenditure classes for rural and urban areas in terms of index numbers where average expenditure for all classes on gur and sugar is set equal to 100 and the expenditure in each class expressed as a percentage thereof.[5]

The manufacture of gur can be divided into the three stages of extraction, clarification and evaporation of the cane juice.[6] The earliest method of crushing consisted of grinding the cane with a pestle and mortar with the pestle moved by bullocks. This method was followed by crushing the cane between two wooden or iron rollers fitted close to one another vertically or horizontally. Such crushers driven by bullocks are still typically employed by small-scale producers. Power-crushers of varying designs and sizes are used by the small number of large-scale manufacturers.[7] Following the process of extraction the juice is clarified and then concentrated by boiling and evaporation. The resultant semi-liquid mass, when cooled, solidifies into gur.

The production of white sugar differs in that it involves the centrifugal separation of sugar crystals from molasses.[8] Crystal sugar is produced by both vacuum pan and open pan sulfitation techniques, the latter being associated with smaller-scale units than the sugar mills proper. The proportion of total output accounted for by open-pan producers is very small (roughly 5 per cent).[9]

Data on labour and capital per unit of output for the two products are taken from Prasad as shown in Table 11.2. Gur thus uses more labour but less capital in relation to output than sugar. The latter is roughly double the price of the former as Table 11.3 shows for rural and urban areas. When the figures for labour productivity in Table 11.2 in terms of tons are adjusted to reflect relative prices, the higher sugar price raises labour productivity in sugar as opposed to gur manufacture. Since the ratio of value-added to gross output is also higher for sugar (25.2 per cent) than gur manufacture (13.4 per cent),[10] labour productivity measured in terms of labour per unit of value added would further increase the differential productivities. With capital productivity the adjustments of course work in the opposite direction though capital per unit value added in gur production (17.7) remains below that for sugar (19.5).[11]

Table 11.1

	Monthly per capita expenditure classes (rupees)													
	0–8	*8–11*	*11–13*	*13–15*	*15–18*	*18–21*	*21–24*	*24–28*	*28–34*	*34–43*	*43–55*	*55–75*	*75+*	*Average*
Rural (all India)														
Numbers in each class as percent	4.39	11.36	9.51	10.77	14.6	12.57	9.2	8.19	7.31	5.82	3.23	1.72	1.33	
Gur	15.6	40.6	49.9	62.5	87.5	90.6	103.1	121.9	153.1	206.3	221.9	340.6	350.0	100
Sugar	3.8	15.4	30.8	38.5	57.7	69.2	96.2	138.5	169.2	238.5	269.2	403.8	980.8	100
Urban (all India)														
Numbers in each class as percent	2.28	4.92	5.14	7.28	10.34	11.39	10.31	9.44	11.63	9.41	7.75	5.18	4.93	
Gur	22.2	61.1	66.7	88.9	77.8	111.1	88.9	105.6	105.6	111.1	172.2	122.2	100	100
Sugar	10.1	17.4	30.4	40.6	53.6	71.0	79.7	94.2	114.5	146.4	168.1	202.9	249.3	100

Source: Indian National Sample Survey (1961–2).

Table 11.2

	Gur	Sugar
Output per person in 100 days	2.1 tons	8 tons
Total productive capital per ton	236.79 rupees	982.00 rupees
Total productive capital per person	503.25 rupees	7859.00 rupees

Source: Prasad (1963), pp. 99–100.

Table 11.3

	Rural	Urban
Gur*	0.48	0.55
Sugar*	1.13	1.06

**Note:* Rupees per seer. A seer is approximately equal to one kilogram.
Source: Calculated from Indian National Sample Survey

THE SIMULATION METHODOLOGY

In the subsequent sections an attempt is made to estimate the employment effects of alternative simulated income redistribution using both aggregative and disaggregative approaches. Before beginning the estimation procedure, however, it is useful to set out the broad simulation methodology which underlies this study. This can be expressed as follows.[12] Given a number of total per capita expenditure classes n, average per capita expenditure in each class $\bar{Y}_i (i = 1, \ldots, n)$, and the percentage of population in each class $\bar{N}_i (i = 1, \ldots, n)$ and $\Sigma \bar{N}_i = 100$) total expenditure can be expressed as:[13]

$$Y = \sum_{i=1}^{n} \bar{Y}_i \bar{N}_i, \tag{11.1}$$

average per capita expenditure as:

$$Y_w = \sum_{i=1}^{n} \frac{\bar{Y}_i \bar{N}_i}{N}, \tag{11.2}$$

where $N = 100$. In the case of equal numbers in each group,

$$Y_w = \sum_{i=1}^{n} \bar{Y}_i \Big/ n,$$

otherwise as here,

$$Y_w \neq \sum_{i=1}^{n} \bar{Y}_i \Big/ n. \qquad (11.3)$$

The share of each group in total expenditure:

$$K_i = \frac{\bar{Y}_i \bar{N}_i}{Y}, \qquad i = 1, \ldots, n. \qquad (11.4)$$

The simulation takes the form of changing the elements of the vector comprising the shares K_i with Y and $\bar{N}_i$ given.[14] As can be seen from (11.4) the result is a new $\bar{Y}$ for each class, i.e. a new level of average per capita expenditure in each class. An estimated consumption function for each commodity concerned is applied to these new average per capita expenditure levels for each class. The summation across the classes yields a new estimated total consumption expenditure on the commodity which may be compared with the pre-redistribution total to arrive at the percentage change.

ESTIMATES OF CONSUMPTION FUNCTIONS

In choosing the best-fitting function applied to the thirteen expenditure classes in Table 11.1 one problem is that there are unequal numbers of persons in the different classes. This presents a problem in estimation inasmuch as the sum of the residuals weighted by these numbers is not zero (though the unweighted residuals do sum to zero).

In order to get the sum of the weighted residuals to sum to zero some form of weighting is required. This was achieved by weighting each of the initial thirteen observations by the numbers in each class, i.e. effectively raising the total number of observations in the proportions $\bar{N}_i (i = 1, \ldots, n)$. The functions estimated on the basis of the weighted regression analysis are shown below:

Gur (rural) $\quad Y = -0.0001x^2 + 0.023x - 0.095, \qquad \bar{R}^2 = 0.976$

$\qquad\qquad (-17.62) \qquad (36.02) \; (-9.05)$

Sugar (rural) $Y = 0.000056x^2 + 0.016x - 0.12, \quad \bar{R}^2 = 0.994$
$(11.56) \quad (30.15)\ (-13.83)$

Gur (urban)[15] (1) $\log Y = -1.32 - \frac{9.18}{x}, \quad \bar{R}^2 = 0.79$

$(-31.17)(-12.6)$

(2) $Y = -0.000042x^2 + 0.0059x + 0.063, \quad \bar{R}^2 = 0.682$
$(-8.08) \quad (9.24) \quad (4.91)$

Sugar (urban) $Y = -0.00018x^2 + 0.037x - 0.18, \quad \bar{R}^2 = 0.995$
$(-34.06) \quad (58.46)(-14.14)$

(*t* values in parentheses)

Since the functions are strictly concave (the coefficient of x^2 is negative) an income redistribution towards the average will in each case increase total consumption except for sugar in rural areas where there will be a decrease.[16] In the case of the alternative log-reciprocal function for urban gur nothing a priori can be said.

THE CONSUMPTION AND EMPLOYMENT EFFECTS OF AN INCOME REDISTRIBUTION

A series of income redistributions have been simulated by changing the shares of total expenditure accruing to the thirteen expenditure classes.[17] Tables 11.4 and 11.5 show the new configurations of income between classes separately for rural and urban areas (with total income given and constant). For each simulation the associated Lorenz ratio is shown at the foot of each table, together with the percentage of total income transferred.

Simulations (1), (2) and (3) are designed to eliminate all those below the absolute poverty line of 15 rupees per month[18] (i.e. the bottom four classes are eliminated), and simulation (7) represents complete equality in the sense of a Lorenz ratio equal to zero.

The tables show that the elimination of absolute poverty in this period would have required a transfer of income representing 10 per cent of the total income in urban areas and 6.2 per cent in rural.[19] The implied reduction in the Lorenz ratio is 32 per cent (rural) and 39 per

Table 11.4 Rural

Expenditure class	*Original share in total expenditure as percentage*	*Simulation* (1)	(2)	(3)	(4)	(5)	(6)	(7)
0–8	1.3	3.0	3.0	3.0	2.0	2.0	3.0	4.4
8–11	5.0	7.8	7.8	7.8	6.0	5.5	5.0	11.4
11–13	5.3	6.6	6.6	6.6	6.0	5.5	5.3	9.5
13–15	6.9	7.4	7.4	7.4	7.0	7.0	6.9	10.8
15–18	11.0	11.0	10.2	11.0	12.0	11.0	11.0	14.6
18–21	11.3	11.3	10.4	11.3	12.0	11.3	11.3	12.6
21–24	9.5	9.5	8.8	9.5	11.0	9.5	9.5	9.2
24–28	9.8	9.0	9.0	9.8	10.0	9.8	9.8	8.2
28–34	10.3	9.5	9.5	9.1	10.0	10.3	10.3	7.3
34–43	10.1	9.1	9.3	8.9	8.0	10.1	10.1	5.8
43–55	7.2	6.2	6.6	5.9	6.0	7.0	7.2	3.2
55–75	5.0	3.7	4.6	3.7	4.0	4.5	5.0	1.7
75+	7.2	5.9	6.8	6.0	6.0	6.5	5.6	1.3
	100	100	100	100	100	100	100	100
Lorenz ratio	0.31	0.21	0.23	0.21	0.25	0.29	0.28	0.0
Percentage of total income transferred		6.2	6.2	6.2	5.9	1.4	1.7	22.4

cent (urban). There is little reason to suppose that these figures would be much altered at the present time if the poverty line were adjusted upwards by an index of prices relevant to the absolute poor.[20]

The consumption effect – disaggregated basis

The percentage changes in consumption for gur and sugar can be estimated as follows: the changed shares in total income of Tables 11.4 and 11.5 imply new average total per capita incomes in each class and the chosen consumption functions are applied to these for each class. Multiplying the outcome by the given (and constant) numbers in each class and summing over all classes yields the estimated new total demands. These are shown in Table 11.6. The last column in Table 11.6 represents the estimates for gur using the alternative log-reciprocal function. They do not differ much from those based on the quadratic function.

Table 11.5 Urban

Expenditure class	*Original share in total expenditure as percentage*	*Simulation*						
		(1)	*(2)*	*(3)*	*(4)*	*(5)*	*(6)*	*(7)*
0–8	0.4	1.7	1.7	1.7	1.2	2.0	2.0	2.3
8–11	1.6	3.6	3.6	3.6	2.5	2.0	1.6	4.9
11–13	2.0	3.7	3.7	3.7	3.0	2.0	2.0	5.1
13–15	3.3	5.3	5.3	5.3	3.9	3.3	3.3	7.3
15–18	5.5	7.5	7.5	7.5	6.0	5.5	5.5	10.3
18–21	7.2	8.3	8.3	8.3	8.0	7.2	7.2	11.4
21–24	7.5	7.5	7.5	7.5	7.9	7.5	7.5	10.3
24–28	7.9	7.9	6.8	7.9	8.2	7.9	7.9	9.4
28–34	11.6	11.6	10.0	9.6	11.8	11.6	11.6	11.6
34–43	11.5	9.9	9.9	9.5	10.0	11.5	11.5	9.4
43–55	12.2	10.4	10.5	10.2	11.5	12.2	12.2	7.7
55–75	10.7	8.6	9.8	8.7	10.0	10.0	10.7	5.2
75+	18.6	14.0	16.0	16.5	16.0	17.3	17.0	4.9
	100	100	100	100	100	100	100	100
Lorenz ratio	0.36	0.2	0.22	0.22	0.28	0.32	0.33	0.0
Percentage of total income transferred		10	10	10	5.6	2.1	1.6	25.8

The table shows an increased demand for gur in both cases which is significantly high for urban areas. The predicted decline in consumption of sugar in rural areas as opposed to its increase for the urban sector reflects the much higher initial consumption in the latter case and the marked shift away from gur in the middle and upper income classes in the rural case. The strict convexity of the function in the case of rural areas (i.e. the rising mpc) is dependent upon the extremely high estimate for the last class (see Table 11.1). Since this figure for the combined sample estimate is based on widely divergent sub-samples[21] and in view of the fact that without this observation the linear (convex) function gives a fractionally better fit[22] it is likely that the predicted decline in consumption is an overstatement.

By taking a weighted average of rural and urban areas the composite total consumption effect can be obtained. The weights used are the shares of rural and urban sectors in the total output of gur and sugar in 1961–2.[23] These combined results are presented in Table 11.7

Table 11.6*

Simulation	*Percentage change in demand for gur*				*Percentage change in demand for sugar*				
	Rural		*Urban*		*Rural*		*Urban*		*Estimates for gur*
(1)	+3.6	(0.22)	+7.9	(0.07)	−2.4	(0.35)	+8.6	(0.18)	+7.8
(2)	+2.0	(0.24)	+5.7	(0.06)	−1.3	(0.39)	+6.3	(0.18)	+6.8
(3)	+3.6	(0.22)	+5.5	(0.06)	−2.4	(0.35)	+5.7	(0.18)	+7.0
(4)	+3.1	(0.27)	+4.6	(0.10)	−2.1	(0.40)	+5.0	(0.26)	+4.9
(5)	+1.5	(0.31)	+2.3	(0.11)	−1.0	(0.47)	+2.5	(0.29)	+2.4
(6)	+2.1	(0.30)	+2.2	(0.12)	−1.4	(0.46)	+2.4	(0.30)	+1.9
(7)	+7.7	(0)	+13.2	(0)	−5.2	(0)	+14.6	(0)	+11.8

Notes: Pre-redistribution specific concentration ratios are 0.51 for sugar and 0.33 for gur in rural areas. In urban districts the figures are 0.33 and 0.12 for sugar and gur respectively. Figures in brackets denote the specific concentration ratios associated with each simulation. They are simply the Lorenz ratios for particular commodities.

Table 11.7

Simulation	*Percentage change in demand for gur*	*Percentage change in demand for sugar*
(1)	+5.0	+5.7
(2)	+3.2	+4.3
(3)	+4.2	+3.6
(7)	+9.5	+9.5

for the absolute poverty simulations (1), (2) and (3) and the totally egalitarian case as these are directly comparable.

The fall in the demand for sugar in rural areas is thus offset by the high share of urban areas while the large rural share in gur consumption pulls down the higher predicted increase in urban consumption.

An alternative set of calculations was performed on the absolute poverty simulations for gur using the discrete data instead of the continuous function and making the *ex ante* mpc assumption.[24] In both rural and urban areas the results were very similar to those shown in Table 11.6.[25] The main difference between the methods lies not in the overall estimates but in the fact that with a continuous (and strictly concave) function any redistribution which is an unambiguous movement towards the average increases consumption while in the discrete case this does not always hold because of the discontinuities in the behaviour of the marginal function.

The employment effect

With the assumption of a constant labour-output ratio, employment increases at the same percentage rate as output. To define the increase in terms of actual numbers, however, requires applying the labour-output ratios for gur and sugar to the estimated incremental output. Taking the output of gur and sugar in 1961–2, calculating the rural and urban shares and applying the percentage increases in demand of the last section yields the incremental output. Multiplying by the labour-output ratios of 7:100 tons for sugar and 47:100[26] tons for gur gives the additional numbers demanded for each simulation. Table 11.8 presents these results for rural and urban areas.

Table 11.8*

Simulation		*Change in demand for labour* *Gur* *Urban*	*Gur* *Rural*	*Sugar* *Urban*	*Sugar* *Rural*	*Gur* *Urban + rural*	*Sugar* *Urban + rural*
(1)	(+71,205)	+72,118	+66,724	+11,473	−1,112	+138,842	+10,361
(2)	(+62,076)	+52,034	+37,068	+8,308	−602	+89,102	+7,706
(3)	(+63,902)	+50,208	+66,724	+7,517	−1,112	+116,932	+6,405
(4)	(+44,731)	+41,992	+57,456	+6,594	−973	+99,448	+5,621
(5)	(+21,909)	+20,996	+27,801	+3,297	−463	+48,797	+2,834
(6)	(+17,344)	+20,083	+38,922	+3,165	−648	+59,005	+2,517
(7)	(+107,721)	+120,501	+142,715	+19,122	−2,409	+263,216	+16,845

**Notes:* Actual numbers. Figures in brackets represent estimates based on the log-reciprocal function for urban gur.

THE AGGREGATIVE APPROACH – THE DIRECTION AND EXTENT OF AGGREGATION BIAS

Unlike the disaggregative approach used above which estimates the change in consumption and employment for gur and sugar separately, the aggregative approach groups the two products together into a single category of sweetener and applies a consumption function estimated for the two products combined.[27] We take first the situation in the rural areas. To ease the computational requirements it was decided to apply the consumption function for the composite good to the changed income levels associated with just one simulation – that of perfect equality with a Lorenz ratio equal to zero. On this basis, consumption of the composite good was found to increase by 3.4 per cent. Since this is equivalent to growth of both gur and sugar at 3.4 per cent the incremental outputs of the two were obtained in the manner described for the disaggregative approach. Labour and capital requirements were then found using the same labour/output and capital/output ratios as in the disaggregated approach. Table 11.9 shows the differential increases in capital and labour predicted by the two approaches on the basis of simulation (7).

The aggregative approach thus understates both capital and labour requirements relative to the approach which considers the shifts between the two products. In the case of labour the understatement is extremely marked.

In the urban sector it is plain from Table 11.6 that there is an almost equi-proportionate rise in the two products so that no aggregation bias results. This result can probably be explained by the fact that sugar typically has a satiety level since there is no (or very little) shift to higher qualities as incomes rise.[28] In their extensive study of English family budgets Prais and Houthakker thus found that of all the commodities studied, only in the case of sugar were no such shifts apparent.[29]

Table 11.9

	Disaggregative approach	*Aggregative approach*	*Aggregation bias*
Increase in labour (actual numbers)	140 306	64 592	54%
Increase in capital (millions of rupees)	53.0	44.1	17%

CONCLUSIONS

This chapter has examined the problem of aggregation bias in the use of existing models which purport to estimate the employment effects of an income redistribution. That is, it has examined for one industry the distortions resulting from the aggregation into a homogeneous group products which in fact are produced by highly disparate methods of production and consumed by different income groups. In the Indian sugar processing industry it was found that while no bias exists for the urban sector, aggregative estimates for the rural areas involve significant distortions. In particular it was found that labour requirements were considerably understated relative to the results reached by the disaggregative approach. Understatement was also found in the case of capital albeit to a lesser extent. Obviously only very limited conclusions can be drawn on the basis of results from a single industry in a single country. However, similar relationships found between different types of products, technologies and incomes in other industries suggest that our results may have wider significance.[30] In any event there is a clear need for further studies on the basis of disaggregated data to test the ubiquity and direction of aggregation bias for various industries and countries.

Notes

1. ILO (1970).
2. These are surveyed in Cline (1975).
3. Morawetz (1974), p. 506.
4. *Indian Sugar Yearbook* (1969–70), p. 119.
5. A method of presentation developed by Mahalanobis (1958). In all cases here and in what follows estimates used are those of the combined sample in the Indian National Sample Survey (1961–2).
6. Gur manufacture is described in Government of India (1965) and Roy (1951).
7. Of the total number of sugar cane crushers in use 96 per cent are worked by bullocks. Calculated from Government of India (1959–60).
8. The techniques of production of white sugar are described in detail in Baron (1975).
9. Baron (1975).
10. Ratios of value-added to gross output were calculated from data given in Prasad (1963).
11. Assuming a price differential of 2:1 between sugar and gur.
12. A similar formulation for equal numbers in each class is to be found in Paukert, Skolka and Maton (1976).

13. Instead of as here regarding the total population as 100, actual numbers can be used so that actual total expenditure is $\sum_{i=1}^{n} \bar{Y}_i \bar{N}_i K$, where K is a scalar.
14. An alternative is to regard $\bar{N}_i$ as variable so that an increase in income for a group is reflected in fewer numbers in the group. Meade characterizes this as 'contraction of the poor' as opposed to 'enrichment'. (See Meade (1976).) Where, however, the distribution of individuals households about the mean income is unknown for each group this procedure is difficult.
15. In view of the relatively inferior fit in the case of urban gur two alternative functions have been presented (and used in the simulations). Strictly speaking the adjusted coefficients of determination ($\bar{R}^2$) of these two functions cannot be compared since the dependent variables (Y and log Y) are different. Goldberger suggests an adjustment procedure to put the functions on the same footing so that $\bar{R}^2$ can be used as the choice criterion in choosing the best fitting function. See Goldberger (1964), p. 217.
16. Strict concavity of a consumption function means that linear interpolation between any two points always gives lower values for consumption than the function itself. An analogous use is made of strict concavity by Sen in the context of utilitarian welfare economics. See Sen (1973).
17. The simulations are based on the assumption that the income redistributions do not alter the price relative between sugar and gur which is given in estimating the functional forms.
18. This is defined at 1960–1 all-India prices as 15 and 20 rupees per month for rural and urban areas respectively. See Bardhan (1974). Dandekar and Rath use a slightly higher figure of 22.5 rupees for urban areas. See Dandekar and Rath (1971). The figure of 21 rupees used here may be regarded as an average of these estimates.
19. Dandekar and Rath regard this figure in absolute terms as a conceptual measure of unemployment and under-employment inasmuch as it represents the amount of additional employment needed to remove absolute poverty. For 1968–9 the figure for rural areas is 822.4 rupees crores. See Dandekar and Rath (1974).
20. Over the period 1958–68 it has been suggested that 'disparities in consumption in real terms increased'. See Chatterjee and Bhattacharya (1974). Bardhan finds that 'between 1960–1961 and 1967–1968 there has been a very significant rise in the percentage of people below the minimum level'. See Bardhan (1974), p. 276.
21. The combined sample estimate is 2.55 rupees and the three subsample estimates are 1.18, 1.62 and 3.89 rupees.
22. Though this is also true of gur in rural areas the sub-sample estimates are far more convergent and hence more reliable. The combined estimate is 1.12 rupees, the average of 1.15, 1.05 and 1.16 rupees.
23. Total production of gur in 1961–2 was 5 885 827 tons, 67 per cent of which was demanded by rural areas. In the same period output of sugar comprised 2 546 000 tons; the urban sector dominating (74 per cent) the consumption in this case. Total production estimates are from the *Indian Sugar Yearbook*, 1969–70. The rural and urban shares are taken from the

average consumption of gur and sugar in volume terms presented in the Indian National Sample Survey 1961–2.

24. The *ex ante* mpc assumption is that for each individual the change in consumption is equal to the change in income multiplied by the pre-redistribution mpc. The difficulty with this assumption is that it makes ΔC insensitive to the size of ΔY. We have used in the simulations the *ex post* apc assumption, i.e. each individual in moving to a new level assumes the apc of those already at that level. With this assumption the mpc is not insensitive to the size of ΔY. Our approach follows Cline (1972).
25. For rural gur the percentage increases in consumption were 3.2, 2.4 and 3.0 per cent for simulations (1), (2) and (3) respectively. For urban gur the corresponding estimates were 8.3, 7.6 and 8.3 per cent.
26. The labour-output ratio for sugar shown in Table 11.2 above, which is based on a 100 day season, has here been reduced to reflect the fact that the season varies between 135 and 200 days. (See Baron (1975).) Gur, on the other hand, operates on the basis of a 100-day season. The capital-output ratio for sugar has also been reduced in the same proportion to a figure of 549.92 from 982 rupees.
27. The estimated consumption function for the rural areas is

$$\underset{(-5.7)}{y = -0.000046x^2} + \underset{(44.28)(-14.92)}{0.039x - 0.22}, \qquad \bar{R}^2 = 0.993$$

(figures in parentheses represent t ratios).

28. Although there is no actual point of satiety in the case of sugar in urban areas, there is nevertheless a marked decline in the mpc as incomes rise.
29. Prais and Houthakker (1955).
30. On the maize-grinding and laundry soap markets in Kenya see respectively Stewart (1977) and Langdon (1975).

References

Bardhan, P.K. (1974). 'On the Incidence of Rural Poverty in the Sixties', in Bardhan, P.K. and Srinivasan, T.N. (eds), *Poverty and Income Distribution in India*, Statistical Publishing Society.

Baron, C. (1975). 'Sugar Processing Techniques in India', in Bhalla, A.S. (ed.), *Technology and Employment in Industry*, ILO, Geneva.

Bhalla, A.S. (ed.) (1975). *Technology and Employment in Industry*, ILO, Geneva.

Chatterjee, G.S. and Bhattacharya, N. (1974). 'On Disparities in Per Capita Household Consumption in India', in Bardhan, P.K. and Srinivasan, T.N. (eds), *Poverty and Income Distribution in India*, Statistical Publishing Society.

Cline, W.R. (1972). *Potential Effects of an Income Redistribution on Economic Growth*, Praeger.

Cline, W.R. (1975). 'Distribution and Development: A Survey of Literature', *Journal of Development Economics*, Vol. 1.
Dandekar, V.M. and Rath, N. (1971). 'Poverty in India – I: Dimensions and Trends', *Economic and Political Weekly*, Jan.
Dandekar, V.M. and Rath, N. (1974). 'Right to Gainful Work,' in Bardhan, P.K. and Srinivasan, T.N. (eds), *Poverty and Income Distribution in India*, Statistical Publishing Society.
Goldberger, A. (1964). *Econometric Theory*, Wiley.
Government of India (1959–60). *Sugar in India*.
Government of India (1965). *Report of the Sugar Enquiry Commission*, Delhi.
Indian National Sample Survey (1961–2). *Consumer Expenditure Survey*, 17th Round.
Indian Sugar Yearbook (1969–70), p. 119.
ILO (1970). *Towards Full Employment*, Geneva.
Langdon, S. (1975). 'Multinational Corporations' Taste Transfer and Underdevelopment: A Case Study from Kenya', *Review of African Political Economy*, Jan.–April.
Mahalanobis, P.C. (1958). 'Science and National Planning,' *Sankhya*, Vol. 17.
Meade, J.E. (1976). *The Just Economy*, Allen & Unwin.
Morawetz, D. (1974). 'Employment Implications of Industrialization in Developing Countries: A Survey', *Economic Journal*, Sept.
Paukert, F., Skolka, J. and Maton, J. (1976). *Redistribution of Income, Patterns of Consumption and Employment*, ILO, Geneva.
Prais, S.J. and Houthakker, H.S. (1955). *The Analysis of Family Budgets*, Cambridge University Press.
Prasad, K. (1963). *Technological Choice Under Development Planning*, Popular Prakashan.
Roy, S.C. (1951). *Monograph on the Gur Industry of India*, Indian Central Sugarcane Committee, Kanpur.
Sen, A.K. (1973). *On Economic Inequality*, Clarendon Press.
Stewart, F. (1977). *Technology and Underdevelopment*, Macmillan.

Appendix 11.1
A Geometrical Representation of the Relationship between Income Redistribution and Consumption*

In this note we present a simple geometrical framework in which it is possible to examine the impact of an income redistribution in favour of the poor on the consumption of a commodity. (In Chapter 11 the commodities concerned were gur, sugar and a composite of the two sweeteners, but the analysis can be generalized.) It is also possible within this framework to investigate the question of which distribution of income produces the maximum consumption of a particular commodity (or, more generally, consumption expenditure as a whole).

In Figure 11A.1 over the range Y_1 Y_2 the curvilinear consumption function C is strictly concave since the function is such that linear interpolation between any two points always gives lower values for C than the function itself, i.e. between Y_1 and Y_2:

$$[SC\,(Y_1) + (1 - S)\,C\,(Y_2)] < \mathrm{C}\,[S\,(Y_1) + (1 - S)\,(Y_2)]$$

$$(\text{where } S < 0 < 1)$$

any averaging of incomes between Y_1 and Y_2 must therefore raise total consumption.[1] In the case of strict convexity, averaging of incomes will reduce total consumption since

$$[SC\,(Y_1) + (1 - S)\,C\,(Y_2)] > C\,[S\,(Y_1) + (1 - S)\,(Y_2]$$

and linear interpolation yields higher values than the function. Where the function is linear the left-hand side of the inequalities is equal to the right-hand side so that averaging produces no effect on consumption.[2]

Using these concepts the process of income redistribution and its effect on consumption can be illustrated by the four-quadrant diagram of Figure 11A.2.

In Figure 11A.2 the lower right-hand quadrant shows a normal distribution (assumed for simplicity) with N representing numbers of individuals at each income level $\bar{Y}$ (where $\bar{Y}$ represents the average total per capita income for each income class to make the analysis correspond to grouped consumption

* From J. James (1976) *Technology, Products and Income Distribution: A Conceptualisation and Application to Sugar Processing in India*, ILO, World Employment Programme Research, Working Paper 61.

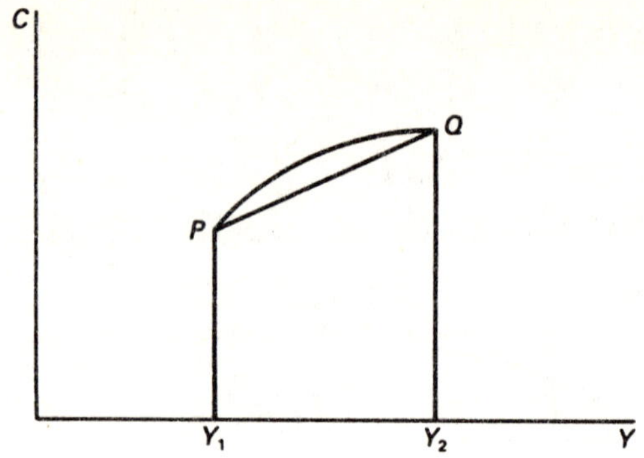

Figure 11A.1

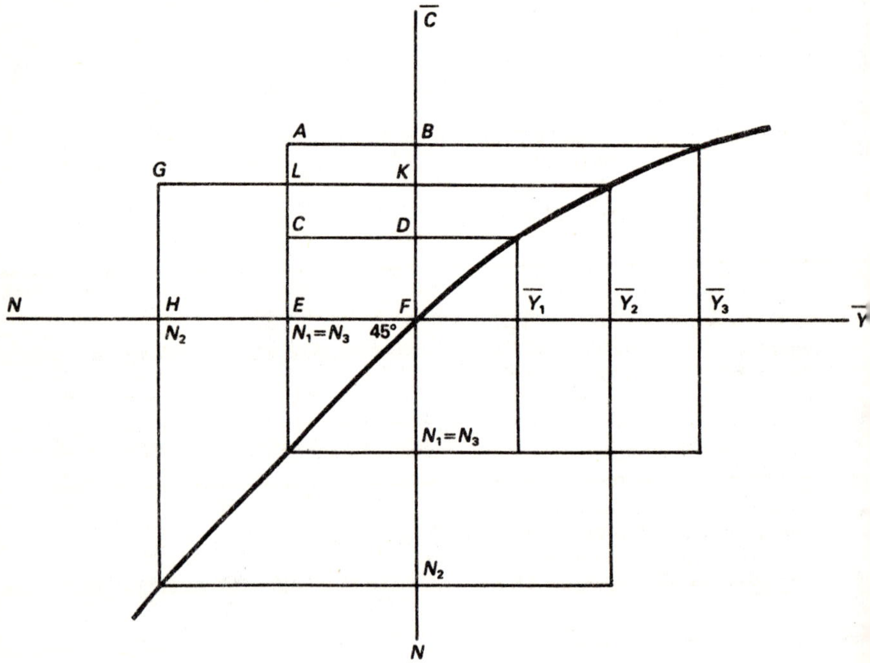

Figure 11A.2

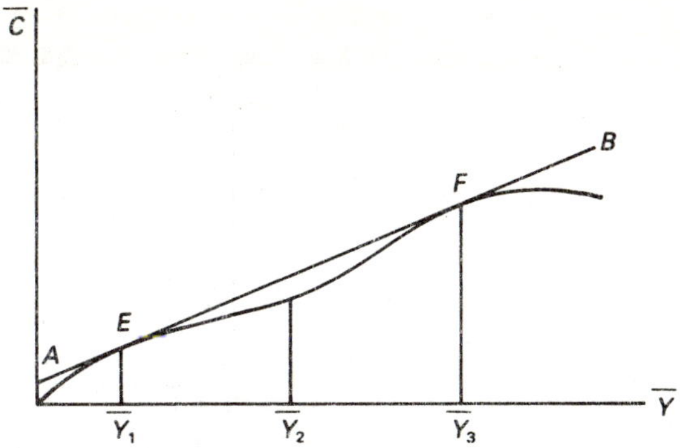

Figure 11A.3

and income data), and $N_1 = \frac{1}{2} N_2$. The upper right-hand quadrant shows the estimated functional relationship between $\bar{Y}$ and $\bar{C}$ (where $\bar{C}$ is the average per capita consumption expenditure for each group). The lower left-hand quadrant merely uses the 45° line to transpose the values of N to the horizontal axis and the last quadrant shows total consumer expenditure, i.e. $\sum_{i=1}^{n} \bar{C}_i \bar{N}_i$.

We assume for the sake of simplicity a community of three income groups so that in the pre-redistribution situation total income (Y) equals $N_1 \bar{Y}_1 + N_3 \bar{Y}_3 + N_2 \bar{Y}_2$ ($N_3 = N_1$). Using the strictly concave consumption function it can be seen in the upper left-hand quadrant that total consumption equals the sum of the rectangular areas *CDEF* + *ABEF* + *GKHF* or $N_1 \bar{C}_1 + N_3 \bar{C}_3 + N_2 \bar{C}_2$.

The society is now made totally egalitarian by transferring $(\bar{Y}_3 - \bar{Y}_2)$ to $\bar{Y}_1$ so that each group has the mean income $\bar{Y}_2$. Total income $2 \times (N_2 \bar{Y}_2)$ is equal to $N_1 \bar{Y}_1 + N_3 \bar{Y}_3 + N_2 \bar{Y}_2$. Total consumption equals $2 \times$ (*GKHF*) and the increase is seen to equal *LKCD* – *ABLK* (since *GLHE* equals *LKEF*). It can also be seen that given the strictly concave consumption function, this configuration of incomes yields the maximum consumption expenditure since, for any movement from the mean income, $ABLK < LKCD$. Analogous but converse results follow for a strictly convex function. With a linear (convex) function no change occurs. With functions not uniformly either strictly convex or concave no a priori conclusions can be drawn. The configuration which maximizes total consumption is, however, derivable.

Along the line segment *AB* of Figure 11A.3 it is true by convexity that

$$\bar{Y}_2 N = S(\bar{Y}_1) N + (1 - S) \bar{Y}_3 N \qquad (0 < S < 1)$$

where $\bar{Y}_2$ is average income and N is the number of individuals with such average income. Maximizing consumption in this case involves notionally moving those with average incomes to $\bar{Y}_1$ and $\bar{Y}_3$ in the proportions S and

Table 11A.1

Name	Form	Slope	Elasticity
1. Linear	$y = a + bx$	b	bx/y
2. Semi-logarithmic	$y = a + b \log x$	b/x	b/y
3. Hyperbolic	$y = a - \frac{b}{x}$	b/x^2	b/xy
4. Double-logarithmic	$\log y = a + b \log x$	by/x	b
5. Log-reciprocal	$\log y = a - \frac{b}{x}$	by/x^2	b/x
6. Quadratic	$y = ax^2 + bx + c$	$2ax + b$	$\frac{2ax + b}{y/x}$

where:

y = expenditure on the specific commodity
x = total income
$\frac{dy}{dx}$ = slope
$\frac{(x\ dy)}{(y\ dx)}$ = elasticity.

$(1 - S)$ (which can be solved from the last equation). This result follows from the fact that below AB consumption can be improved by, as it were, shifting those at the mean to the local maxima E and F. Should redistribution have as its aim the maximum consumption of a commodity(ies) some would have to be made poorer and some richer if the curve were shaped as in Figure 11A.3 with strict convexity and concavity over part of the income range.

We may now examine the properties of the widely used Engel curves in order to discover their implications for consumption when used in simulation exercises. Table 11A.1 sets out these curves in terms of derivatives and elasticities.

We may classify these by the behaviour of the slope (mpc).

1. *MPC falls monotonically*:
   (a) semi-logarithmic function
   (b) hyperbola
   (c) quadratic function with coefficient of x^2 negative
   (d) double-logarithmic with $b > 0 < 1$.
2. *MPC rises monotonically*:
   (a) quadratic function with coefficient of x^2 positive
   (b) double-logarithmic with $b > 1$.
3. *MPC constant*:
   (a) linear function.
4. *MPC rises (falls) and then falls (rises)*:
   (a) log-reciprocal.

Notes

1. In terms of Lorenz curve analysis it may be noted that only a lower Lorenz ratio with non-intersecting curves constitutes an unambiguous move towards the average, i.e. a redistribution which moves everyone towards the average or some towards the average and none away from it.
2. An analogous use is made of strict concavity by Sen in the context of utilitarian welfare economics showing how, with a strictly concave welfare function, averaging raises social welfare. (A.K. Sen (1973). *On Economic Inequality*, Clarendon Press.)

Appendix 11.2
A Note on the Historical Relationship Between Products, Processes and Incomes

The aggregation bias that was revealed in Chapter 11 is not the result of pure chance; on the contrary, it reflects a systematic set of underlying patterns in the relationship between products, processes (technologies) and incomes in developing countries. The purpose of this short note is to examine the nature of these patterns and thereby to provide a basis on which the results reported in Chapter 11 might be generalized.

The point of departure is the recognition that developing countries are characterized in general by the coexistence of forms of production which originate in very different historical periods.[1] For one thing, much of the production structure of these economies contains modes of production which belong to the period of 'pre-capitalist economic formations'. In this period, as Marx put it, 'the greater part of the products are produced for the satisfaction of the immediate needs of the community not as commodities',[2] in contrast to the later capitalist mode of production which turns 'the commodities formerly produced as immediate use-values into exchange values'.[3] The origin of the historical relationship between products, processes and incomes may thus be said to reside in communities where the mode of production exists essentially to provide for the subsistence needs of the community. That these earlier modes of production continue to survive in developing countries reflects the fact that (especially in Africa and parts of Asia) vast numbers still live at subsistence levels of income and thus demand the type of commodity which is appropriate to the earlier historical period. For the most part, this means commodities containing a high proportion of essential characteristics that are produced by self-employed and other small-scale producers in and for impoverished groups of households typically in rural areas.

The organized or formal sector of the developing economy, on the other hand, operates mostly on the basis of imported processes of production developed in and for the incomes and product requirements of advanced economies. These requirements are generally such that product designs come to acquire an increasingly high proportion of 'high-income' (including labour- and time-saving) characteristics;[4] partly as a consequence, the techniques required to produce them become increasingly capital-intensive and large scale. (One important dimension of product quality, for example, is based on the need for 'close tolerances' and it is often very difficult to achieve these without relatively sophisticated capital equipment.[5] What evolves consequently is a close relationship over time between products (defined in terms of embodied characteristics) and the nature of production processes (with both being related to changes in the level of income per head).

Broadly speaking, therefore, the relationships between products, processes and incomes that can currently be observed in a developing country (that is, cross-sectionally) are in fact the reflection of what has occurred historically, ranging all the way from pre-capitalistic economic formations to the most modern products and processes that, frequently, are associated with the activities of multinationals from the advanced countries. (Indeed, in the context of one sector, cotton textiles in India, it has been argued that a continuum exists: that is, with rising incomes consumers seem to pass through the product and process characteristics of the earliest pre-capitalist formations to those of successively later periods.[6])

Plausible though they may seem, these relationships are not always recognized in the literature; in fact, many of those who have addressed the question can find no theoretical basis for the patterns that have just been identified (See, for example, Cline, Cooper and Morawetz.[7]) The last-mentioned author, for example, can see no 'a priori reason' why an association between products (defined in terms of 'high' and 'low-income' characteristics) and factor proportions 'should be generally true'.[8]

Yet, on balance the evidence presented in the earlier chapters tends to refute this sceptical view. The case studies cited in Chapter 2 in particular (as well as those contained in the appendix to that chapter), mostly show the relationships between products, processes and incomes that one, on historical grounds, would expect. The evidence presented in Chapter 3 also supports this historically based view, for it shows that in only a minority of cases do multinationals from advanced economies adapt products in any significant way to the specific needs and circumstances of the majority of those who live in poor countries.

Notes

1. This paragraph is based on James (1976). The argument is extended in James (1977).
2. Marx (1946), p. 377.
3. Marx (1964), p. 116.
4. For a detailed discussion see Stewart (1977).
5. Forsyth, McBain and Solomon (1982).
6. James (1976).
7. Cline (1975), Cooper (1990) and Morawetz (1974).
8. Morawetz (1974), p. 514.

References

Cline, W.R. (1975). 'Employment Implications of Industrialization in Developing Countries: A Survey', *Journal of Development Economics*, Feb.

Cooper, C. (1990). 'Notes on Income Distribution, Growth and Demand', in Stewart, F., Thomas, H. and de Wilde, T. (eds), *The Other Policy*, IT Publications.

Forsyth, D., McBain, N. and Solomon, R. (1982). 'Technical Rigidity and Appropriate Technology in Less Developed Countries', in Stewart, F. and James, J. (eds), *The Economics of New Technology in Developing Countries*, Pinter.

James, J. (1976). 'Products, Processes and Incomes: Cotton Clothing in India', *World Development*, Vol. 4 (2).

James, J. (1977). *Technology, Products and Income Distribution: A Conceptualisation and Application to Sugar Processing in India*, ILO, World Employment Programme Research, Working Paper No. 61.

Marx, K. (1946). *Capital*, Allen & Urwin.

Marx, K. (1964). 'Pre-Capitalist Economic Formations', in Hobsbawm, E.J. (ed.), *Pre-Capitalist Economic Formations*, Lawrence & Wishart.

Morawetz, D. (1974). 'Employment Implications of Industrialization in Developing Countries: A Survey', *Economic Journal*, Sept.

Stewart, F. (1977). *Technology and Underdevelopment*, Macmillan.

Index